AF541111

INTERMEDIATE FLUID MECHANICS

INTERMEDIATE FLUID MECHANICS

BY

WILLOW HUGHES
DENAYE DAWSON

KNOWLEDGE BAKERS

ISBN: 9789390013272 (HB)

Published by : **Knowledge Bakers**
Pune, India
Email: knowledgebakers2019@gmail.com

Digitally Printed at : **Replika Press Pvt. Ltd.**

PREFACE

Fluid mechanics is the study of fluid behavior (liquids, gases, blood, and plasmas) at rest and in motion. Fluid mechanics has a wide range of applications in mechanical and chemical engineering, in biological systems, and in astrophysics. Intermediate Fluid Mechanics is a level of study in fluid mechanics that is more advanced than introductory courses, but less advanced than graduate-level courses. It is typically taken by undergraduate students who have already completed an introductory course in fluid mechanics and mathematics. In Intermediate Fluid Mechanics, students learn more advanced topics in fluid mechanics such as potential flow, viscous flows, boundary layer flows, and turbulence effects. The mathematical tools used in fluid mechanics, such as vector calculus and differential equations, are also covered in greater depth. The goal of Intermediate Fluid Mechanics is to provide students with a deeper understanding of fluid mechanics that is necessary for pursuing a degree in engineering or physics. It also prepares students for graduate-level study in fluid mechanics and for careers in industries such as aerospace, automotive, and energy.

The book starts with an introduction to fluid mechanics and its importance in various fields. It also discusses the types of fluids, their properties, and the basic principles of fluid mechanics. It focuses on the mathematical tools required for understanding fluid mechanics and covers vector calculus, tensor analysis, and differential equations. It introduces the panel method, which is a numerical technique used to solve potential flow problems and discusses the Bernoulli equation and its applications and also focuses on the basics of potential flow, which is an idealized model for fluid flow that assumes the fluid is inviscid and incompressible. It covers viscous flows, which are flows that involve frictional forces due to the viscosity of the fluid and focuses on boundary layer flows, which are a type of viscous flow that occurs near solid surfaces. It covers laminar and turbulent boundary layers and their characteristics. It introduces integral boundary layer relationships, and covers the statistical description of turbulence and some basic turbulence models. The book is primarily intended for undergraduate students who have already taken introductory courses in fluid mechanics and mathematics. It provides a more advanced treatment of fluid mechanics that is suitable for students pursuing a degree in engineering or physics.

We are grateful to all those persons as well as various books, manuals, periodicals, magazines, journals etc. that helped in the preparation of this book. In spite of the best efforts, it is possible that some errors may have occurred into the compilation and editing of the book. Further queries, constructive suggestions and criticisms for the improvement of the book are always welcomed and shall be thankfully acknowledged.

Willow Hughes
Denaye Dawson

CONTENTS

I. Introduction

This book is meant to be a second course in fluid mechanics that stresses applications dealing with external potential flows and intermediate viscous flows. Students are expected to have some background in some of the fundamental concepts of the definition of a fluid, hydrostatics, use of control volume conservation principles, initial exposure to the Navier-Stokes equations, and some elements of flow kinematics, such as streamlines and vorticity. It is not meant to be an in-depth study of potential flow or viscous flow, but is meant to expose students to additional analysis techniques for both of these categories of flows. We will see applications to aerodynamics, with analysis methods able to determine forces on arbitrary bodies. We will also examine some of the exact solutions of the Navier-Stokes equations based on classical fluid mechanics. Finally we will explore the complexities of turbulent flows and how for boundary layer flows one can predict drag forces. This compilation is drafted from notes used in the course Intermediate Fluid Mechanics, offered to seniors and first year graduate students who have a background in mechanical engineering or a closely related area.

In developing some of these more advanced topics there will be a number of mathematical tools developed and applied to specific flow situations. An early introduction to some of the basic concepts is presented in Chapter 2. But other mathematical tools and manipulations are introduced later as the topics require. Much of fluid mechanics can be developed from a mathematical point of view and students should realize that much of the early development was from mathematicians, such as Bernoulli, Euler, Navier, Stokes, whose names should sound familiar, and many others. However, as presented here the physical interpretations and applications are important and an attempt is made to develop analysis methods with an understanding of the physical consequences along side of all of the mathematical constraints and requirements of a problem or situation. It is hoped that the student not only learns the equations and how to manipulate them but also to understand the physical situations and how the physical flow phenomena are interpreted.

Fluid mechanics, as its name implies, is a subset of the larger field of mechanics. Mechanics is a branch of science dealing with forces and motion, and their relationships. Mechanics has static and dynamic elements. Forces may exist without motion and/or with motion and forces may initiate or change motion. Since fluid properties are significantly different than solids, fluid response to applied forces can be much more complex and difficult to describe. Due to fluid deformation rates (yielding to forces over time) fluids have complex distributions of pressure and velocity and acceleration. This spatial distribution of fluid motion is an important part of the understanding of how forces are transmitted within fluids. A major area of study in fluid mechanics is the kinematic motion of the fluid and how this is described. The dynamic flow of fluids is governed by Newton's law of momentum conservation whereby forces are required to accompany a rate change of momentum. Since forces and motion are all around us and influence much of what we do in our everyday lives it is not surprising that the origins of mechanics, and fluid mechanics, dates back to the ancient Greeks. Archimedes was instrumental in developing the concepts of hydrostatics which are used to understand forces by fluids on its surroundings as well as how fluid pressure changes due to gravitational forces. Much of the interesting applications of fluid motion follow the formulations of the conservation of momentum. In this sense one is interested in how fluid motion is altered by the existence of imposed forces on the fluid. This is important in the applications of fluid transport (pipelines, biological systems, lubrication, chemical reactions and a host of other applications). It is also important when objects move within fluids. Since we are surrounded by fluids in our living environment, by either air or water, any motion of an object must deal with the fact that fluids must be "pushed out of the way". That is to say, object motion translates into fluid motion. And based on Newton's Law of reaction, forces acting on objects by fluids are related to forces by objects acting on fluids. So the guiding principles of fluid motion analysis comes from conservation principles of momentum when tied to other constraints of conservation of mass and energy.

The conservation equations illustrate the physical relationship governing fluid flow and the resultant forces on and by fluid flow. Flows are often classified by the inclusion of certain forces and/or certain effects. The largest classification is most likely between inviscid and viscous flows. This is a major division used in the development here. We first present inviscid flows where we analyze forces either on or by fluids and how they affect fluid motion. The primary forces are caused by pressure distributions and gravity. As we will see the pressure field or distribution in space, is influenced by the fluid motion. Consequently, the resulting forces can become rather complicated as pressure and velocity are intertwined. In contrast, gravity represents a constant force proportional to the mass of fluid as it plays a role in affecting fluid motion. After examining some of the inviscid flow situations, we introduce viscous effects and discuss how viscous dominated flows are of importance and how they may be analyzed. As we shall see the approach to these flows is very different because of the physical conditions, boundary conditions and resultant analysis methods that are used. The coupling of viscous effects, pressure and velocity create complex flow dynamics.

Potential flows are irrotational and allow for the determination of the flow field and pressure field based on the use of a scalar velocity potential. Potential flows ignore frictional effects. This has many advantages mathematically as well as allows for a good physical interpretation of the flow. However, there is lost information concerning any viscous effects that provide additional forces that may alter the flow field and pressure distribution. The underlying assumption is that these effects are minor for certain types of flows. One may hear different classifications of flows such as potential flow, ideal flow, incompressible or constant property flows. Potential flow, as mentioned, allows the replacement of the velocity vector with a velocity potential, which is a scalar and proves mathematically useful for many situations. We will deal with potential flows as inviscid, incompressible and also irrotational. The latter condition is that the vorticity is zero throughout the flow (except maybe at some singularity points within the flow). The definition of the velocity potential mathematically requires flows to be irrotational, as we shall see. The consequence of this is that the viscous terms vanish and the expression used for the acceleration of the fluid can be simplified. The solution to the resultant governing equations becomes very much simplified.

Another class of flows, known as ideal flows, are inviscid and incompressible. Typically, incompressible flows are those that do not have (significant) changes in density with changes of pressure. Liquids tend to be incompressible except under extreme conditions of high pressure changes. Gases may or may not be compressible depending on how large any pressure changes may be within the flow. Since pressure changes are linked to velocity changes within a flow it is possible to classify compressible effects through the flow conditions. Note that variations of density within a fluid flow can be caused by temperature effects, or if the fluid has a variation in concentration of some solute, but the flow still be treated as incompressible.

When including viscous effects the additional force may work with or against other forces, such as those due to pressure variations or body forces such as gravity. Interestingly the inclusion of viscous forces requires a model that depends on the properties of the fluid. This model relates how friction is measured within a moving, deforming fluid. It is not a universal law, say like conservation of mass or energy, it is mostly an empirical relationship that has been shown to be valid for a wide range of fluids and conditions. However, one can expect exceptions to this for some "exotic" fluids. We introduce the viscosity as a property of the fluid which defines the needed viscous force that results in a specified fluid deformation rate. This type of model is based on the work in the 1900s by Navier and Stokes who, separately, developed a brilliant formulation to account for viscous effects. This formulation is developed from the Cauchy equation which itself includes viscous effects in an overall identification of forces, or stresses, acting within the fluid flow field. The model allows the evaluation of stress terms based on deformation experienced by the fluid caused by viscous forces. In our applications we will restrict analyses to incompressible, constant property flows to be able to assess the contributions of the various terms. This has wide reaching applications, but does not delve into the realm of gas dynamics which combines certain equations of states with the conservation of mass, momentum and energy to evaluate compressible flows.

Another very important aspect of viscous forces is the fluid-surface boundary condition. Fluids, in contact with a solid surface, will tend to have the "no-slip" boundary condition. This is stated mathematically as the velocity at the interface is equal to the velocity of the surface, in both magnitude and direction. Moving away from this boundary the fluid velocity will change and the rate of change is found to be related to the frictional force on the surface caused by the fluid. By Newton's third law there is an equal and opposite force by the surface on the fluid. One must treat this boundary condition based on the view of the fluid as a "continuum". That is to say one does not get down to the molecular level, since at that scale molecules of fluid are bouncing off and on to the surface. The scale at which a fluid can be thought of as a continuum is larger than the mean free path of the molecules of the fluid, which is an average distance molecules travel before they collide with other molecules. For a continuum, the fluid velocity at the surface takes on the value of the surface velocity. There are applications, such in very small scale flows (sub-micron scale) where one can not treat the fluid as a continuum and in these situations "slip" may occur. Also, for rarified gases, under very low pressure, molecules can be very far apart and a continuum is not a valid approach. This transitions into the realm of the kinetic theory of gases. It is beyond the scope of this course to deal with these conditions.

Fluid properties are an important part of fluid flow analyses. Numerical results obviously greatly depend on values of say, density, viscosity, and other properties like surface tension or compressibility. However, here we do not go into these effects specifically. We do note that density is important in the relationship between pressure and velocity. This is easily noted by placing your hand out of the window of a car and comparing the resultant pressure on your hand with that of sticking it out of a boat into the water below when moving at the same speed as in the car. Since water density is nearly one thousand times greater for water compared with air, so is the resultant pressure. Also, a highly viscous fluid like honey will have different flow characteristics than a fluid which a much lower viscosity like water, as seen when pouring honey versus water from a jar under the action of the gravitational force. We will attempt to retain fluid properties in problems that are discussed so these types of distinctions become obvious. However, we will not go into the details of conditions of highly variable properties and the resultant change in flow and forces caused by this variability.

The rest of this book is organized as follows. Chapter 2 develops most of the common mathematical tools required for the rest of the book, although not exclusively. This is to provide a common level for mathematical notation and some manipulation. Chapter 3 develops a generalized Bernoulli equation, useful for later sections in the book. It also helps to explain the less restrictive conditions on the equation compared to that experienced by most first course students. Chapters 4 and 5 develop potential flow methods and solutions and Chapter 6 utilizes this approach in developing the Panel Method for solving for pressure and forces on objects in external flow. Chapters 7, 8 9 and 10 all deal with viscous flows. Chapter 7 develops the Navier-Stokes equation and chapter 8 provides a few classical "exact" solutions. Chapters 8 develops the boundary layer equations, the Blasius solution, and chapter 9 develops the integral solution method. Finally, Chapter 10 explores turbulence, its basic physics, some scaling conditions and a brief application to boundary layer flows. After completion of this book, students should have a better understanding of how to analyze both potential flows and viscous flows for incompressible conditions.

On-line material

- Potential Flow: MIT: http://web.mit.edu/2.016/www/handouts/2005Reading4.pdf
- Efluids for general material and examples and images and videos: http: //www.efluids.com
- APS gallery of motion: http://gfm.aps.org

II. Mathematical Tools

In this chapter we introduce a few mathematical tools that we will use in formulating some of the analysis of fluid flow problems for both inviscid and viscous flows. We introduce the use of tensor notation which is widely used in expressing fluid mechanics governing equations. We also show some mathematical manipulations that will help to provide some physical insight into the governing equations.

Tensor Notation

Most students are very familiar with vector notation (or Gibbs notation) for describing (usually) three component vectors in fluid mechanics. Vector notation implies the existence of components of the vector. A scalar, as is known is fully described by a single number, its magnitude. Of course a vector quantity may have more than three components in general, but here we are using vectors to describe components in three dimensional space. Each component may be a function of both space and time, and as such the entire vector is a function of space and time. The number of components in the vector is determined by the "dimensionality" of the vector. For instance a velocity field may only have two orthogonal directions, say (x,y), and as such the flow becomes two dimensional – the assumption is that there is no velocity variation in the third, z, direction. We can think of the three components as a set of three independent variables that the variable in question is a function of.

Tensor notation is an alternative approach and is a very powerful way of expressing any dimensional vector, as well as what are known as higher order tensors – variables that have several sets of independent variables to be considered. We say that a scalar is a zero order tensor and a vector is a first order tensor, such as velocity. An example of a second order tensor is stress. It requires two sets of indices to define its local value. A third order tensor requires a set of three indices to be fully defined. In our three dimensional world then, a first order tensor has three components. A second order tensor has nine components, determined by the possible combinations of the three elements within each set of indices. To make this clear note that for some vector V in three dimensions:

$$\mathbf{V} = u\mathbf{i} + v\mathbf{j} + w\mathbf{k} \; = u_i$$

Where u,v, and w are the values of the three components each a function of (x,y,z), $\mathbf{i}, \mathbf{j}$ and $\mathbf{k}$ are the unit vectors associated with x,y and z coordinates respectively and u_i is tensor notation for a vector V. The tensor representation uses a subscript which can have three possible values, each representing one of the three components the vector, or $u_1 = u, u_2$ = v and $u_3 = w$. Each component is denoted by a particular subscript that is defined based on the numerical value of the index. Here index number "1" is representative of the x coordinate, etc. But this is not limited to Cartesian coordinates. For example subscripts 1, 2 and 3 could represent r, θ and z in cylindrical coordinates.

Also, we can define a stress tensor as:

τ_{ij}

and the number of possible combinations of the subscripts i,j are nine. Do not confuse these subscripts with the unit vectors since they can take on one of the possible three directions. For fluid flow problems we will define the stress tensor and use it to arrive at our conservation of momentum equation. A third order tensor could be written as T_{ijk} and it would have 27 elements (3x3x3). As stated above, a single subscript denotes a vector and has 3 possible values, one for each coordinate.

It is possible to write all of our vector equations using tensors instead of the vector notation. This has some advantages and is used widely in fluid mechanics. Table 2.1 illustrates a number of examples of the equivalence of vector and tensor notation. A few of these are worth discussing and we will be using them as we write out the various equations that we will be formulating and using. It is suggested that you examine this notation and look at the various operators listed.

Operations among vectors carry with them some special consideration and rules. The gradient operator is ∇ in vector notation. If the quantity of interest is the gradient of scalar "a", written as ∇a , then this operation results in a vector whose components are partial derivatives in each of the three coordinates, $\frac{\partial a}{\partial x}$ is the x component, etc. In tensor operation this is expressed as $\frac{\partial a}{\partial x_i}$ which is a vector whose components correspond to each of the three possible values of the index "i". If the gradient is being taken of a vector, say u_j, the the resulting gradient expression is a second order tensor with the two indices, i for the partial derivatives and j for the vector quantity u_j. This is then written as $\frac{\partial u_j}{\partial x_i}$. Notice that the gradient operator index comes first since it operates on the vector u_j. There are nine combinations of i, j for this second order tensor. We will see this tensor in our discussions of viscous flows and vorticity.

Table 2.1 Listing of a few vector, tensors or scalars in vector and tensor notation

	Examples of Vector VS Tensor Notation	
Quality	Vector Notation	Tensor Notation
Scalar	a	a
Vector	$\bar{a}$	a_i
2nd order tensor	$\bar{\bar{a}}$	a_{ij}
dot product (scalar)	$\bar{a} \cdot \bar{b}$	$a_i b_i$
cross product (vector)	$\bar{a} \times \bar{b}$	$\varepsilon_{ijk} a_j b_k$
del operator (vector)	$\bigtriangledown$	$\frac{\partial}{\partial x_i}$
gradient (vector; tensor)	$\bigtriangledown a; \bigtriangledown \bar{a}$	$\frac{\partial a}{\partial x_i}; \frac{\partial a_i}{\partial x_j}$
divergence (scalar)	$\bigtriangledown \cdot \bar{a}$	$\frac{\partial a_i}{\partial x_i}$
curl (vector)	$\bigtriangledown \times \bar{a}$	$\varepsilon_{ijk} \frac{\partial a_k}{\partial x_j}$
Laplace operator (scalar)	$\bigtriangledown^2$	$\frac{\partial^2}{\partial x_i \partial x_i}$
divergence of a 2nd order tensor	$\bigtriangledown \cdot \bar{\bar{a}}$	$\frac{\partial a_y}{\partial x_i}(vector)$
dot product: vector & del	$\bar{a} \cdot \bigtriangledown$	$a_j \frac{\partial}{\partial x_j}(scalar)$
dot product: vector & gradient	$\bar{a} \cdot \bigtriangledown \bar{b}$	$a_j \frac{\partial b_i}{\partial x_j}(vector)$
Material Derivative	$\frac{\partial}{\partial t} + \bar{u} \cdot \bigtriangledown or \frac{D}{Dt}$	

Vorticity: (pseudovector)	$\omega_k = \nabla \mathrm{x} \bar{u} = \left(\frac{\partial u_i}{\partial u_j} - \frac{\partial x_j}{\partial x_i}\right) = \varepsilon_{ijk}\frac{\partial u_k}{\partial x_j}$	
Rate of strain tensor: (2nd order tensor)	$\frac{1}{2}\left(\frac{\partial u_i}{\partial u_j} + \frac{\partial u_j}{\partial u_i}\right)$	

Consider now the divergence operation, in vector notation written as $\nabla \cdot V$. This operation is the dot, or scalar, product between the vector operator ∇ and vector V. The result is a scalar. The tensor notation for this operation is $\frac{\partial u_i}{\partial x_i}$ where the indices for the vector u_i are the same as the index for the partial derivative operator. This implies that the same component of each vector, ∇ and V, should be used simultaneously. The rule in tensor notation then is to add all of these component operations to result in one final scalar. This becomes for the three coordinates, x_1, x_2 and x_3, which in Cartesian coordinates are x, y and z: $\frac{\partial u_i}{\partial x_i} = \frac{\partial u_1}{\partial x_1} + \frac{\partial u_2}{\partial x_2} + \frac{\partial u_3}{\partial x_3} = \frac{\partial u}{\partial x} + \frac{\partial v}{\partial y} + \frac{\partial w}{\partial z}$

The result is a scalar formed from the sum of each of the partial derivatives. As is apparent from this equation, when a term has a "repeated" index (such as the index "i" in the above equation for both u and x) then the rule is to sum over all values of the index. This applies to a particular term in an equation.As an additional example consider the term $\frac{\partial a_{ij}}{\partial x_i}$ in this case the index "i" is repeated and j is left as a "free index" (which could take on any of the possible three coordinate values.) Consequently this expression is a vector, with one free index, whose components are represented by the value of "j".There is also an operator, ε_{ijk}. This is given the name "permutation operator". Notice in Table 2.1 this is listed as used in the cross product and curl operation (the curl is the cross product of the vector operator ∇, which in tensor notation is written as $\frac{\partial}{\partial x_j}$, and the vector u_k). The permutation operator definition results in the cross product between two vectors. For this each of the indices can take on three different values, and there are 27 different possible combinations. The rules associated with the different values are as follows:

- if any of the three indices have the same value then $\varepsilon_{ijk} = 0$
- if the order of the indices is cyclic (1,2,3 or 2,3,1 or 3,1,2) then $\varepsilon_{ijk} = 1$
- if the order of the indices is anti-cyclic (1,3,2 or 2,1,3 or 3,2,1) then $\varepsilon_{ijk} = -1$

The result is three values for the permutation operator, with values being +1, -1, or 0. Consequently,

the following occurs for the cross product: a_j x b_k
$a_j \times b_k = \varepsilon_{ijk} a_j b_k = (a_2 b_3 - a_3 b_2)\,\boldsymbol{i} + (a_3 b_1 - a_1 b_3)\,\boldsymbol{j} + (a_1 b_2 - a_2 b_1)\,\boldsymbol{k}$

The reader should perform this operation for the curl of a vector where a_i is replaced by $\frac{\partial}{\partial x_j}$ and b_k is replaced by u_k

There is another often used operator in tensor notation called the Kronecker delta, δ_{ij}. The rule used is that when $i = j$ the value of δ_{ij} is one, otherwise it is set equal to zero, that is:

$$\delta_{ij} = 1 \; for \; i = j$$

$$\delta_{ij} = 0 \; for \; i \neq j$$

This comes in handy when one wants to include terms where there may be a nonzero value only when $i = j$. We will see this in our expression in the Navier-Stokes equation for the pressure contribution to the total force on a fluid element. For example suppose we want to include a gradient of a scalar, S, as $\frac{\partial S}{\partial x_j}$ in an equation, note that this is a vector quantity. However, if we are evaluating different components of the gradient we only want to include those components that align with the vector component "i", then we would write this term $\frac{\partial S}{\partial x_j}\delta_{ij}$. In doing this we see that when we evaluate this term we only include the gradient component in the i direction, since when operating on the gradient with δ_{ij} results in a nonzero possible value only when $i = j$. We will use this for the pressure term in the Navier-Stokes equation as is explained later. We will use gradients, divergence and curl operations in the expressions used in the Navier-Stokes equations. The reader should become familiar with these.

Gradient, Divergence and Curl Operators

The gradient is a vector operator that for instance, when operated on a scalar produces a vector. It takes the partial derivative with respect to each component of the coordinate system. It is denoted as $\bigtriangledown$ and can be expressed in Cartesian coordinates as:

$$\bigtriangledown() = \frac{\partial()}{\partial x} i + \frac{\partial()}{\partial y} j + \frac{\partial()}{\partial z} k$$

Where a scalar of a vector or tensor could be inserted into the parenthesis. A gradient of a vector then has 9 components, where you take x,y,z derivatives of each of the three vector components. This would form the secord order tensor:

$$\bigtriangledown V = \begin{bmatrix} \frac{\partial u}{\partial x} & \frac{\partial u}{\partial y} & \frac{\partial u}{\partial z} \\ \frac{\partial v}{\partial x} & \frac{\partial v}{\partial y} & \frac{\partial v}{\partial z} \\ \frac{\partial w}{\partial x} & \frac{\partial w}{\partial y} & \frac{\partial w}{\partial z} \end{bmatrix}$$

If one takes the dot product of $\bigtriangledown$ with a vector we get the divergence of that vector which is a scaler:

$$\bigtriangledown \cdot V = \frac{\partial u}{\partial x} + \frac{\partial v}{\partial y} + \frac{\partial w}{\partial z}$$

Note that the divergence of the second order tensor yields a vector, such as $\bigtriangledown \cdot \bigtriangledown V$. The x component of this vector is:

$$\bigtriangledown \cdot \bigtriangledown u = \left(\frac{\partial^2 u}{\partial x^2} + \frac{\partial^2 u}{\partial y^2} + \frac{\partial^2 u}{\partial z^2} \right) i$$

The y and z components should be apparent where the x component of velocity is replaced by the y and z components of velocity in the numerators for the y and z components of the resultant vector.

Vector Identities

Vector Identity

Below are shown a few vector identities that are used in fluid mechanics in the manipulation of the governing equations of motion. A,B are vectors, phi, is a scalar.

$$\bigtriangledown (A \cdot B) = (A \cdot \bigtriangledown) B + (B \cdot \bigtriangledown) A + A \cdot (\bigtriangledown \cdot B) + B \cdot (\bigtriangledown \cdot A) = A \bigtriangledown B + B \bigtriangledown A$$

$$\frac{1}{2} \bigtriangledown (A \cdot A) = (A \cdot \bigtriangledown) A + A \cdot (\bigtriangledown + A) = A \bigtriangledown A$$

$$\bigtriangledown \cdot (\bigtriangledown \phi) = 0$$

$$\bigtriangledown \cdot (\bigtriangledown \cdot A) = 0$$

$$\bigtriangledown^2 \phi = \bigtriangledown \cdot (\bigtriangledown \phi)$$

$$\bigtriangledown \cdot (\bigtriangledown \cdot A) = \bigtriangledown (\bigtriangledown \cdot A) - \bigtriangledown^2 A$$

Divergence Theorem: What is the name of this symbol? $B \cdot dA = \int\int\int (\bigtriangledown \cdot B)\, dV$ (where dA is an elemental area around the boundary of volume V)

Stokes' Theorem: $\oint B \cdot dl = \int\int (\bigtriangledown \cdot B) \cdot dA$ (where dl is an element of the line integral around the boundary of area A).

There are a number of vector identities and manipulations that are very useful in fluid mechanics. We list a few of these here for future reference. Unless stated otherwise these are valid for any vector quantity, V.

The reader may want to visit the web site: https://en.wikipedia.org/wiki/Vector_calculus_identities

Material Derivative

The material derivative is a shorthand notation expressing the rate change of some property or parameter, which could be a scalar or a tensor, such as temperature, pressure or momentum (mass times velocity). This time derivative is based on a Lagrangian frame of reference of a fixed mass of fluid. So in a sense one is following a fixed mass of fluid and describing how the given property changes. Since the fluid mass can move we say that the fixed mass depends on both space and time. For example if we pick the scalar pressure, P, and say that our fixed mass has a pressure that is a function of space and time using Cartesian coordinates we have for the variable pressure:

$$P = f(x, y, z, t)$$

Then if we want its time rate of change for a fixed mass at a given time we use the notation that DP is the change of quantity P and DP/Dt is its change over time accounting for changes in both space and time (note we apply the chain rule for each of the possible coordinate directions:

$$\frac{DP}{Dt} = \frac{\partial P}{\partial t} + \frac{\partial P}{\partial x}\frac{dx}{dt} + \frac{\partial P}{\partial y}\frac{dy}{dt} + \frac{\partial P}{\partial z}\frac{dz}{dt}$$

Noting that $\frac{dx}{dt} = u$ and similarly for each coordinate direction, the equation becomes:

$$\frac{DP}{Dt} = \frac{\partial P}{\partial t} + u\frac{\partial P}{\partial x} + v\frac{\partial P}{\partial y} + w\frac{\partial P}{\partial z} \qquad (2.1)$$

Things to pay attention to in this equation are (i) the velocity components u, v and w are not vector quantities, but components of the velocity vector, V; (ii) the time variation is specified by the first term and is evaluated at a particular point in space (as are all of the other terms) but all other terms may in fact be time dependent; this equation is valid for scalars or tensors, and the tensor sense is determined by the quantity being evaluated (in Equation (2.1) it is a scalar since P is a scalar). The material derivative of the velocity vector V, is:

$$\frac{DV}{dt} = \frac{\partial V}{\partial t} + u\frac{\partial V}{\partial x} + v\frac{\partial V}{\partial y} + w\frac{\partial V}{\partial z} \tag{2.2}$$

Since V is a vector then DV/Dt is also a vector, and each term on the right hand side is a vector and the components of each of these vectors is determined by inserting the component for the vector V. In other words the x component is:

$$\frac{Du}{Dt} = \frac{\partial u}{\partial t} + u\frac{\partial u}{\partial x} + v\frac{\partial u}{\partial y} + w\frac{\partial u}{\partial z} \tag{2.3}$$

The above is for Cartesian coordinates. This can be written for any selected coordinate system. For example the result in cylindrical coordinates in r, θ is shown below.

Material Derivative in Cylindrical Coordinates

We would like to find an expression for DV/DT in cylindrical coordinates that we can use to help interpret streamline coordinates. We will only examine a two dimensional situation, r, θ since z is similar to Cartesian coordinates.

Given the vector $V = u_r\hat{r} + u_\theta\hat{\theta}$ we write this as a material derivative:

$$\frac{DV}{Dt} = \frac{\partial V}{\partial t} + u_r\frac{\partial V}{\partial r} + \frac{u_\theta}{r}\frac{\partial V}{\partial \theta}$$

When the angular position changes then there can be a change in both the unit vectors $\hat{r}$ and $\hat{\theta}$ since their orientation changes. So we can write the following:

$$\frac{\partial V}{\partial \theta} = \frac{\partial\left(u_r\hat{r} + u_\theta\hat{\theta}\right)}{\partial \theta} = \hat{r}\frac{\partial u_r}{\partial \theta} + u_r\frac{\partial \hat{r}}{\partial \theta} + \hat{\theta}\frac{\partial u_\theta}{\partial \theta} + u_\theta\frac{\partial \hat{\theta}}{\partial \theta}$$

Now we must interpret the derivatives of the unit vectors with respect to θ. Refer to the figure below showing a blob of fluid moving $\partial\theta$ that results in a change of the unit vector $\hat{r}$, that we can write as: $d\hat{r} = |\hat{r}|d\theta\hat{\theta} = d\theta\hat{\theta}$, which we can write as: $\frac{\partial \hat{r}}{\partial \theta} = \hat{\theta}$.

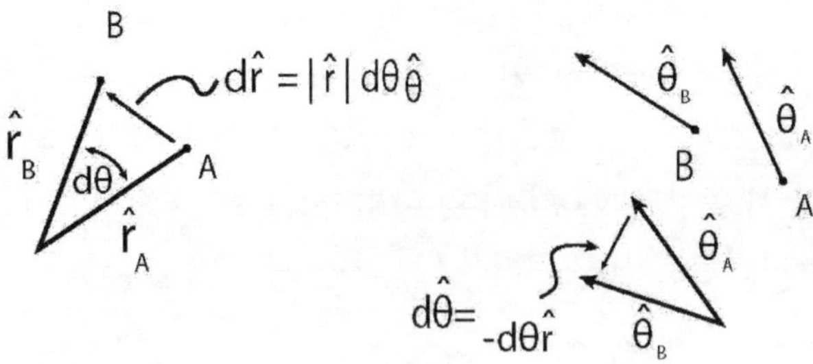

Similarly the change in the unit vector, $d\hat{\theta} = -r d\theta$, as illustrated above on the right as well. This is negative because if the change in the unit vector is counterclockwise the change in the r unit vector is in the negative r direction. The result is $\frac{\partial \hat{\theta}}{\partial \theta} = -r$.

All of this can now be combined and used in the material derivative above resulting in:

$$\frac{DV}{Dt} = \frac{\partial V}{\partial t} + u_r \frac{\partial V}{\partial r} + \frac{u_\theta \partial V}{r \partial \theta} = \left(\frac{\partial u_r}{\partial t} + u_r \frac{\partial u_r}{\partial r} + \frac{\partial_\theta \partial u_r}{r \partial \theta} - \frac{u_\theta^2}{r} \right) \hat{r} + \left(\frac{\partial u_\theta}{\partial t} + u_r \frac{\partial u_\theta}{\partial r} + \frac{u_\theta \partial u_\theta}{r \partial \theta} + \frac{u_r u_\theta}{r} \right) \hat{\theta}$$

This applies to Cylindrical Coordinates, but as we shall see helps to interpret streamline coordinates when there is curvature to the streamline. The net result is added terms that account for the change of the unit vectors due to the curving nature of the flow.

Acceleration

If one finds the material derivative of the velocity vector of a fixed mass of fluid, namely Eqn. 2.2, the result is the acceleration of that fixed mass of fluid in time and space. The general expression for this vector quantity can be written in tensor notation as:

$$\frac{Du_i}{Dt} = \frac{\partial u_i}{\partial t} + u_j \frac{\partial u_i}{\partial x_j} \tag{2.4}$$

Note here that the vector, V is replaced with the tensor notation for a vector written as u_i. Also the summation rule has been used in the second term on the right hand side, this implies that there are really three terms being represented by the one term shown. The first term on the right hand side is denoted as the "local acceleration" and physically represents the rate change of velocity at a fixed point in space. The second expression on the right hand side is denoted as the "convective acceleration" and physically represents how the velocity field changes over space. If a flow is steady the local acceleration is by definition zero. However a fluid mass as it moves through space where the velocity changes in space will experience acceleration (or deceleration). An example is steady flow through a nozzle, or diffuser, where the fluid velocity will increase, or decrease, along the flow direction. For these two cases a fixed mass of fluid will accelerate, or decelerate due to moving in space. The reader is reminded that the components of the acceleration depend on the component of velocity in the derivatives.The convective acceleration can be rewritten using a vector identity listed above. Namely, the convective acceleration in vector notation is $(V \cdot \nabla) V$ and can be expressed as: $(V \cdot \nabla) V = \frac{1}{2}\nabla (\mathrm{V}\cdot\mathrm{V}) - \mathrm{V}\times(\nabla\times\mathrm{V})$. In tensor notation this becomes:

$$u_j \frac{\partial u_i}{\partial x_j} = \frac{1}{2}\frac{\partial (u_j u_j)}{\partial x_j} - \varepsilon_{ijk}\left(u_j \varepsilon_{klm}\frac{\partial u_m}{\partial x_l}\right) \tag{2.5}$$

Consequently, the convective acceleration can be recast as the expression on the right hand side. Although this may not seem like it is of any benefit and just makes things more complicated we will show that it does indeed have advantages. The reader should pay attention to the subscripts in the above expression. Notice that each term is a vector with the free index of "i", all other indices are repeated and therefore summed. In the summation of vectors we basically need to sum components, this implies that they have the same index, in this case "i".

Streamline Coordinates

Streamline Coordinates define the flow direction of the velocity vector. It is defined as the locus of points tangent to the velocity vector at some instant in time. If the flow is steady these lines do not change. For our purposes we will consider steady flow here, but that is not necessary. Fig. 2.1 illustrates the streamline as well as a streamline coordinate system we will define. Our discussion for simplicity will be two dimensional such that flow is within the $s - n$ plane. We retain an orthogonal coordinate system with "s" along the streamline and "n" normal to the streamline. The third coordinate would be normal to both s and n, into the page.

If we take the cylindrical coordinate system as an example and consider flow along a curve, with some non-infinite radius of curvature, then we expect additional terms to appear in the acceleration. These terms account for the changing unit vectors as discussed in the material derivative in cylindrical coordinates. We make note of the fact that the velocity in the n direction is zero, by definition of the streamline. However, there can still be acceleration in the n direction since the unit vectors are changing. Using the cylindrical coordinate expression as a guide and denoting r as normal to the streamline direction, n, and θ as along the streamline direction, s, then the results for acceleration in the s and n directions are:

$$a_s = \frac{\partial u_s}{\partial t} + u_s \frac{\partial u_s}{\partial s} = \frac{\partial u_s}{\partial t} + \frac{\partial {u_s}^2}{2} \tag{2.6}$$

$$a_n = \frac{{u_s}^2}{R}$$

Here R is the local value of the radius of curvature of the streamline. The value of R goes to infinity for a straight streamline and there is no acceleration in the n direction. Recall here that u_r (or u_n) is identically zero at the streamline and that the expression for a_n only contains the last term associated with the changing unit vector θ, or in this case s. Also we define in the streamline coordinates the direction n to be inward towards the radius of curvature as shown in Fig. 2.1. This is opposite to the convention used in cylindrical coordinates where r is radially outward, hence the sign is positive for the value of a_n. That is, if there is a curving streamline the acceleration is inward towards the direction of the center of the radius of curvature.

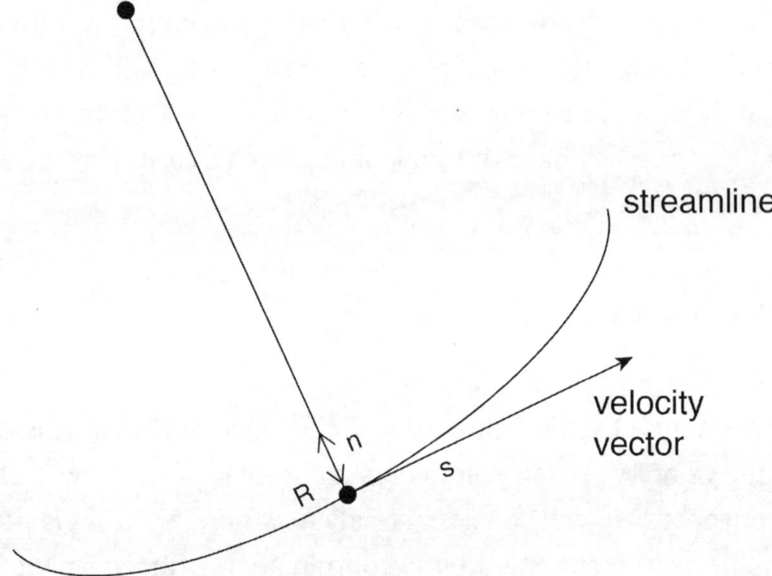

Fig. 2.1 Streamline coordinate illustration; s is along the velocity vector direction and tangent to the streamline and n is normal to s while directed towards the center of radius, denoted as R.

Velocity Gradient Tensor

The velocity gradient tensor describes how the velocity varies near a specified location within the flow field. It is represented as $\frac{\partial u_j}{\partial x_i}$ which is a second order tensor, and therefor has nine components in three dimensional space. A second order tensor can be decomposed into symmetric and antisymmetric parts given as follows:

$$\frac{\partial u_j}{\partial x_i} = \frac{1}{2}\left(\frac{\partial u_j}{\partial x_i} + \frac{\partial u_i}{\partial x_j}\right) + \frac{1}{2}\left(\frac{\partial u_j}{\partial x_i} - \frac{\partial u_i}{\partial x_j}\right) \tag{2.7}$$

Notice here that the first term adds half of the transpose of the original tensor and the second term subtracts the same quantity completing this identity. The first term is symmetric that is by inverting the i and j subscripts there is no change in the value of the element. The second term is antisymmetric such that inverting the subscripts the value of the element has the opposite sign. The ½ multiplier is included in the definition of the symmetric and antisymmetric parts such that their sum results in the original tensor.

In fluid mechanics applications the velocity gradient tensor is a measure of the velocity changes experienced in the fluid flow as one moves away (infinitesimally) from a given point. That is to say this tensor is a point function, having a value at any given point in the flow. We refer to that as a local function in space. It may be time dependent as well. So we can say that this tensor has a symmetric part and an antisymmetric part. As we will see when we discuss the viscous forces within a fluid the symmetric part is representative of the deformation rate, or strain rate, experienced by a fluid element. The antisymmetric part represents the rotation rate of a fluid element. If one examines the curl operator on u_j given previously as $\varepsilon_{kij}\frac{\partial u_j}{\partial x_i}$, where the subscripts indicate that this is the kth element, it can be shown by writing out the terms of the curl operation on u_j that this is equivalent to twice the antisymmetric part of $\frac{\partial u_j}{\partial x_i}$ (or the part in parentheses of the second term in Eqn. (2.7)). Therefore the rotation part of the velocity gradient tensor is one half the vorticity of the flow. To be clear, the antisymmetric part of $\frac{\partial u_j}{\partial x_i}$ is a second order tensor, but from observation if i=j then the value of this is identically zero (all zero components along the diagonal of the tensor). The off diagonal terms are six in total but those with inverted indices are the negatives of the noninverted components ($\frac{\partial u_j}{\partial x_i} = -\frac{\partial u_i}{\partial x_j}$). Consequently there are only three independent values, so we identify this second order tensor as a pseudo-vector, having three components. It is a so called pseudo-vector since the sign is arbitrary. In other words do we set $\frac{\partial u_1}{\partial x_2}$ as a positive or negative quantity? The sign convention is typically selected so that a counter clockwise rotation is a positive value. This means that for

$\dfrac{\partial u_1}{\partial x_2}$ being positive then as u_1 increases in the positive x_2 direction then the flow will tend to rotate clockwise and is a negative rotation direction. This is illustrated in Fig. 2.2. Note that if one considers a negative value of $\dfrac{\partial u_1}{\partial x_2}$ coupled with a positive value of $\dfrac{\partial u_2}{\partial x_1}$ such that

$$\omega_3 = \left(\frac{\partial u_2}{\partial x_1} - \frac{\partial u_1}{\partial x_2}\right) > 0 \tag{2.8}$$

then rotation (or vorticity) about the 3 axis (out of the page in Fig. 2.2) is counterclockwise. Obviously, the signs of the two components of the velocity gradient that go into any of the vorticity components can be positive and/or negative. If in the above example for Eqn. 2.8 the magnitudes of $\dfrac{\partial u_2}{\partial x_1}$ and $\dfrac{\partial u_1}{\partial x_2}$ are equal and their signs are the same then the vorticity component ω_3 will be zero. One can think of the net vorticity or rotation rate vector component to be the combined value from both velocity derivative components, which contribute to rotation about an axis. The same interpretation of the combined effects of the two velocity gradient elements can be made for the strain rate, or the symmetric part of the velocity gradient tensor. When the rotation rates of the vertical and horizontal axes are identical (both elements causing, say, counterclockwise rotation at the same rate) then the strain rate will be zero (the two elements in the symmetric part cancel) and the flow is in pure rotation, as a solid body.

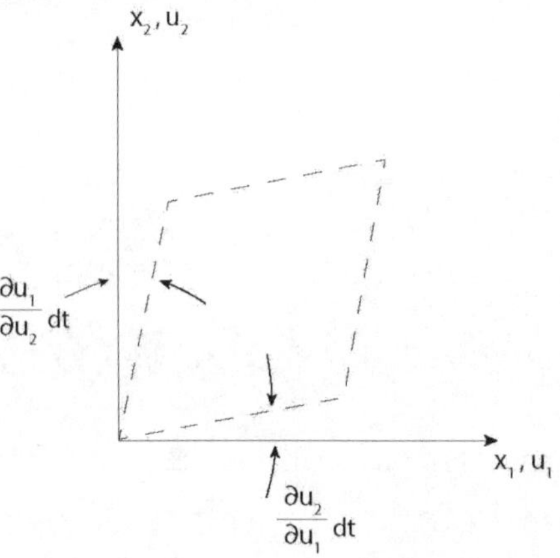

Fig. 2.2 Illustration of rotation rate as determined by the velocity gradient tensor components; the original square element has its sides experiencing rotation at a rate of $\dfrac{\partial u_2}{\partial x_1}$ *for the horizontal axis and* $\dfrac{\partial u_1}{\partial x_2}$ *for the vertical axis.*

Basic Equations

The basic equations generally used in inviscid fluid mechanics are shown in Fig. 2.3. The viscous terms we will add later. But this chart shows various forms of the governing momentum equation which include pressure and gravitational forces contributing to the overall acceleration of the fluid. At the top is the basic Cauchy form of the momentum equation, where viscous effects would be contained in the stress term, τ. The inclusion of viscous terms will be developed in a later chapter. At this point the stress only contains the pressure forces, which is compressive and normal. The streamline coordinate formulation is shown on the right where integration is along the streamline. The introduction of the variable B is a consequence of using the last of the vector identities listed above as equation (2.5) for the convective acceleration term, and then combining the terms shown in the bottom box on the left. Here B is essentially the Bernoulli constant for steady, compressible flow. This is discussed further in the development of the Bernoulli equation. At this point the equations are presented for reference later on when the various terms are developed.

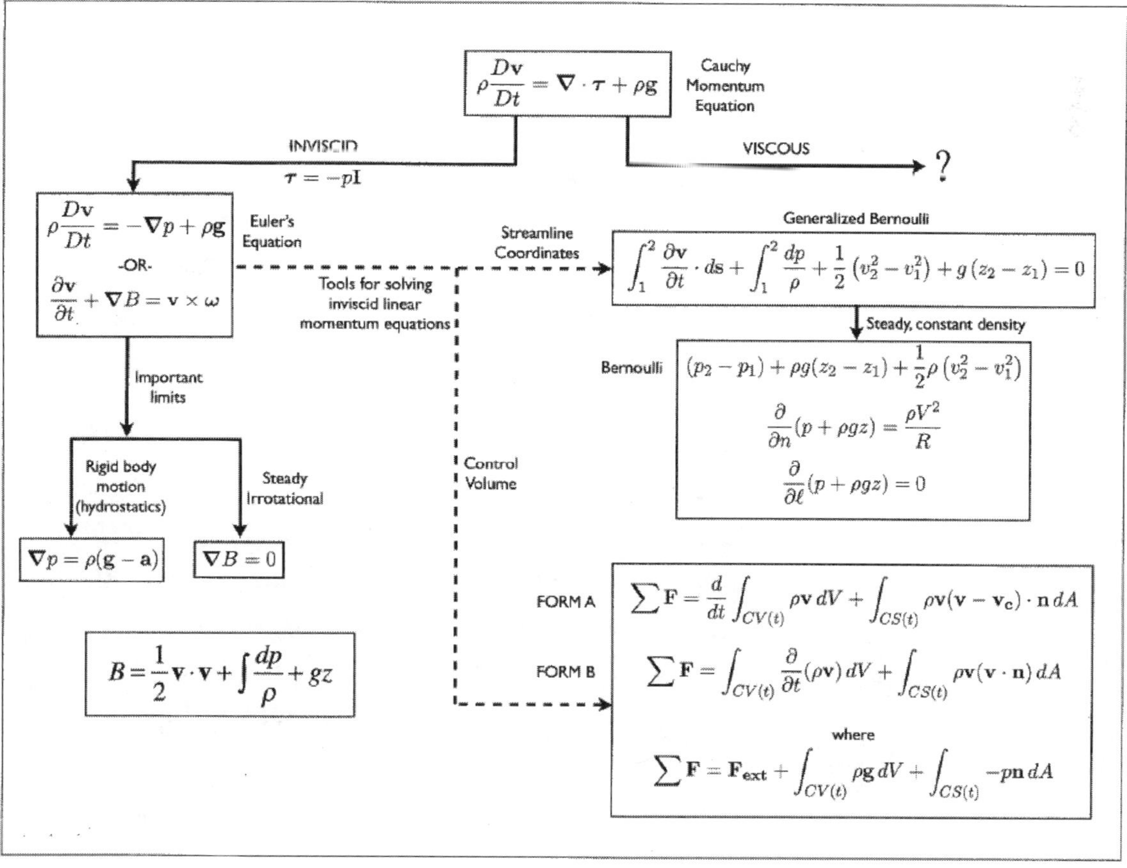

Fig 2.3 Chart Illustrating the basic governing equation in inviscid flow.

III. Bernoulli Equation

Generalized Form

The Bernoulli Equation is presented to most all engineering students and even high school students in a simplified form. This allows the development of a basic understanding of fundamental relationships between velocity and pressure within a flow field. It is typically written in the following form:

$$\frac{P}{\rho} + \frac{V^2}{2} + gz = constant \tag{3.1}$$

The restrictions placed on the application of this equation are rather limiting, but still this form of the equation is very powerful and can be applied to a large number of applications. But since it is so restrictive care must be take in its application. The restrictiof3.ns can be stated as:

- incompressible flow (density is constant)
- viscous forces are assumed to be negligible (no internal fluid friction)
- steady flow (no time dependence)
- flow is along a streamline (apply between two points on the same streamline)

The first condition usually means it can be applied to a liquid or to a gas that has a relatively low velocity such that large changes in pressure do not occur. The second condition is a bit vague at best but assumes other forces, such as caused by pressure changes or body forces are much larger than frictional forces which may be valid when the viscosity is low, and/or when spatial velocity derivatives are not too large. The third term is self explanatory and disallows local acceleration. The last term requires the identification of a streamline and that the evaluation of Eqn. (3.1) occurs between two points along the streamline. A streamline by definition, and as is stated in Chapter 2, assures that the velocity vector is tangent to the streamline. The Bernoulli equation can also be expressed by saying that the constant in the equation is the same at the starting and ending point such that the three terms sum to the same value at these two points and as such can be set equal to each other. Most students reading this will have a fairly extensive use of Eqn. (3.1) and there are many examples that can be found on the internet. The derivation of this equation is also available in many introductory fluid mechanics textbooks. The basic relationship stems from applying the momentum equation (without viscous forces included) along a streamline. This derivation will not be repeated here.

We wish to develop a more general form of the Bernoulli Equation that eliminates the restrictions to incompressible, steady flow along a streamline. The only restriction then is that viscous forces are ignored. The starting point for this development is the differential Euler's Equation for the motion of a fluid element that relates the acceleration to the forces caused by pressure and gravity. This can be expressed using the material derivative from Chapter 2 as a balance of acceleration with pressure and body forces per unit mass of fluid:

$$\frac{D\boldsymbol{V}}{dt}=-\frac{1}{\rho}\frac{\partial P}{\partial \boldsymbol{x}}+\boldsymbol{g} \tag{3.2}$$

Or in tensor notation as:

$$\frac{\partial u_i}{\partial t}+u_j\frac{\partial u_i}{\partial x_j}=-\frac{1}{\rho}\frac{\partial P}{\partial x_i}+g_i \tag{3.3}$$

The pressure term is representative of the net force caused by the compressive load of pressure along the direction of the vector component of interest, on a per mass basis. The body force term is again given as the "i" vector component of the gravitational vector. The acceleration terms on the left can be recast using the vector identity for the convective acceleration given in Chapter 2.

$$u_j\frac{\partial u_i}{\partial x_j}=\frac{1}{2}\frac{\partial\left(u_j u_j\right)}{\partial x_j}-\varepsilon_{ijk}\left(u_j\varepsilon_{klm}\frac{\partial u_m}{\partial x_l}\right) \tag{3.4}$$

The last term introduces the vorticity into the equation since the vorticity is defined as the curl of the velocity vector:

$$\omega_k=\varepsilon_{klm}\frac{\partial u_m}{\partial x_l} \tag{3.5}$$

So the last term becomes the cross product of velocity and vorticity, or:

$$\varepsilon_{ijk}\left(u_j\varepsilon_{klm}\frac{\partial u_m}{\partial x_l}\right)=\varepsilon_{ijk}u_j\omega_k$$

Notice that the result of this operation is a vector in the "i" direction, consistent with the other terms in the Euler equation. By introducing the vorticity into the convective acceleration term, this term can now can be considered to have two components. One is half of the gradient of the magnitude of the velocity vector squared (the first term), and the other is the cross product between the velocity vector and vorticity (a vector). This latter term has been identified as the "Lamb vector", after the applied mathematician, Horace Lamb (1849-1934).

Euler's equation can now be rewritten as:

$$\frac{\partial u_i}{\partial t} + \frac{1}{2}\frac{\partial\left(u_j u_j\right)}{\partial x_i} - \varepsilon_{ijk} u_j \omega_k = -\frac{1}{\rho}\frac{\partial P}{\partial x_i} + g_i \tag{3.6}$$

Next we write the Lamb vector as a gradient of some unknown function:

$$\frac{\partial \pi}{\partial x_i} = \varepsilon_{ijk} u_j \omega_k \tag{3.7}$$

In this expression π must be a scalar quantity such that its gradient is a vector. This is done for convenience so that we can then integrate each term of our equation along any desired path, with elemental distance ds. That is we take the projection of each term along ds (take the dot product of each term with vector ds). However, before we do this we make the following modification to the gravitational term such that we can integrate spatially along ds.

We can write the gravitational term as a "potential" as:

$$g_i = -g\frac{\partial h}{\partial x_i} \tag{3.8}$$

where "h" is a scalar and g is the magnitude acceleration of gravity. The choice of the symbol "h" is because it will have units of length and if gravity is vertically downward then "h" is vertically upward and represents elevation above some arbitrarily chosen datum. Notice that if "i" is horizontal then $\frac{\partial h}{\partial x_i}$ is the change in elevation over a differential change in the horizontal direction, consequently the value of $\frac{\partial h}{\partial x_i}$ is zero. If "i" is in the vertical upward direction then $\frac{\partial h}{\partial x_i} = 1$ and $g_i = -g$. If "i" is at some angle θ to the horizontal, then $\frac{\partial h}{\partial x_i} = \sin\theta$.

We are now able to write the integral along ds of Euler's equation. Note that the dot product of each term results in a scalar that defines the magnitude of the change of the term along direction ds:

$$\int_s \frac{\partial u_i}{\partial t} ds_i + \int_s \frac{1}{2}\frac{\partial\left(u_j u_j\right)}{\partial x_j} ds_i + \int_s \frac{\partial \pi}{\partial x_i} ds_i + \int_s \frac{1}{\rho}\frac{\partial P}{\partial x_i} ds_i + \int_s g\frac{\partial h}{\partial x_i} ds_i = f(t) \tag{3.9}$$

Since the integration is in space there may be a time dependence of each term that is not accounted for which is why the term on the right hand side appears – a general function of time can be added such that if one takes the spatial derivatives this function will vanish. Note that if there is no time dependence then $f(t) = 0$, as well as the first term on the left.

Bringing the time derivative outside of the integral in the first term, and performing the integration along the arbitrary line ds:

$$\frac{\partial}{\partial t}\int_s u_i ds_i + \Delta\left(\frac{(u_j u_j)}{2}\right) + \Delta\pi + \int_s \frac{1}{\rho}\frac{\partial P}{\partial x_i} ds_i + \Delta(gh) = f(t) \tag{3.10}$$

where Δ represents the change of the variable, or expression, along the line s. As shown we have two integrals that we can not evaluate at this point. The first represents the local acceleration and how it varies along the integration path. The second is the pressure term. Note however if the flow is *incompressible*, ρ is constant, then we can say that:

$$\int_S \frac{1}{\rho}\frac{\partial P}{\partial x_i} ds_i = \frac{1}{\rho}\int_s \frac{\partial P}{\partial x_i} ds_i = \Delta\frac{P}{\rho}$$

Equation (3.10) excludes frictional or viscous forces, but that is about its only limitation. As such it is a general form of the Bernoulli Equation. But considering incompressible and steady flow the result is:

$$\Delta\left(\frac{(u_j u_j)}{2}\right) + \Delta\pi + \Delta\frac{P}{\rho} + \Delta(gh) = 0 \tag{3.11}$$

Consequently, the sum of these four terms which represent changes along any direction s is zero, or

$$\frac{(u_j u_j)}{2} + \pi + \frac{P}{\rho} + (gh) = constant \tag{3.12}$$

To satisfy some curiosity, one would expect that by applying the four conditions listed above for the specialized form of Eqn. (3.10) will result in Eqn. (3.1). If density is constant the pressure integral term becomes P/ρ. If the flow is steady then the time dependence terms are zero. If the integration is taken along a streamline then we can make the following argument. The velocity vector by definition is aligned with the streamline, therefore the direction of ds and u_i are identical. The vorticity may be in any arbitrary direction, but the cross product of u_i and ω_j must be perpendicular to u_i. Consequently, since the integration is for the projection of each term along ds, and the cross product of u_i and ω_j is normal to ds (since it is normal to u_i) then the net effect of this term is zero (has no component along ds if s is a streamline). So if integration is along a streamline we can delete the Lamb vector effect, $\Delta\pi = 0$. Combining all of these conditions we end up with Eqn (3.1) as we hoped.

Other than along a streamline, another way in which $\Delta\pi = 0$ is if the vorticity is zero along any chosen integration path. That is to say, the vorticity is zero throughout the flow field. Let's examine the consequences of this condition. If each component of vorticity is zero then we can write a set of conditions, one for each component of the vorticity,

$$\omega_i = \left(\frac{\partial u_k}{\partial x_j} - \frac{\partial u_j}{\partial x_k}\right) = 0; \text{where } i \neq j \neq k \tag{3.13}$$

Now we define a "velocity potential" such that

$$u_j = \frac{\partial \phi}{\partial x_j} \text{ for all values of } j \tag{3.14}$$

This is a rather powerful condition — that a single scalar function,ϕ, can be used to define the velocity field through its partial derivatives. If one takes the derivatives of u_k and u_j as shown in Eqn. (3.13) but rewrite this in terms of the velocity potential, the result is an identity for $j \neq k$. That is to say, it is always = 0 since it is possible to reverse the order of differentiation. Therefore it is concluded that if one can replace the velocity vector with the derivatives of the scalar velocity potential defined in Eqn. (3.14) then the vorticity is zero everywhere. This can also be stated as: a scalar velocity potential exists and can be used to define the velocity field if the flow is vorticity free. Since vorticity can be defined as the degree of local rotation occurring in the flow we say the flow is "irrotational" if the vorticity is zero everywhere.

Using the above definition of a velocity potential, that exists for irrotational flow, it can be said that irrotational flow results in a simplified form of the Bernoulli Equation since π is zero when the vorticity is zero.

Here we summarize the Bernoulli Equation and how it is modified for different conditions.

1. General form
$$\int_s \frac{\partial u_i}{\partial t} ds_i + \Delta\left(\frac{(u_j u_j)}{2}\right) + \Delta\pi + \int_s \frac{1}{\rho}\frac{\partial P}{\partial x_i} ds_i + \Delta\left(gh\right) = f(t)$$
2. Incompressible flow form
$$\int_s \frac{\partial u_i}{\partial t} ds_i + \Delta\left(\frac{(u_j u_j)}{2}\right) + \Delta\pi + \Delta\frac{P}{\rho} + \Delta\left(gh\right) = f(t)$$
3. Steady form
$$\Delta\left(\frac{(u_j u_j)}{2}\right) + \Delta\pi + \int_s \frac{1}{\rho}\frac{\partial P}{\partial x_i} ds_i + \Delta\left(gh\right) = 0$$
4. Irrotational, or along a streamline, form
$$\int_s \frac{\partial u_i}{\partial t} ds_i + \Delta\left(\frac{(u_j u_j)}{2}\right) + \int_s \frac{1}{\rho}\frac{\partial P}{\partial x_i} ds_i + \Delta\left(gh\right) = f(t)$$
5. Combination, steady, incompressible along a streamline, or irrotational

$$\Delta\left(\frac{(u_j u_j)}{2}\right) + \Delta\frac{P}{\rho} + \Delta\left(gh\right) = 0$$

The last form is that which is most often introduced as a first exposure to the Bernoulli equation, yet it does come with a number of conditions, and one must be reminded that all of the above forms exclude any viscous, or frictional, force effects.

It is important to realize that the Bernoulli equation can be used for rotational or irrotational flow, but the former requires that it be applied along a streamline and viscous forces are not included. The later condition of irrotational flow, also neglects viscous forces but assumes irrotational flow so can be applied between any two points within the flow (not necessarily along a streamline).

In the chapters dealing with irrotational flow we will apply the Bernoulli equation between points where we know information, like far upstream of some object when there is flow over the object, to some point where we would like to calculate information, like on the surface of the object. However, since viscous forces are not included care must be taken to not apply the typical viscous boundary condition, of no-slip (forcing the fluid velocity to be equal to the surface boundary velocity.) Consequently inviscid flows allow slip, which means that the surface velocity of the fluid is some value that may need to be determined. This determination is the subject of the next two chapters.

IV. Potential Flow Basics

Potential Flow Basics

Potential flows are those flow situations were the flow is taken to be irrotational, such that the vorticity is zero throughout the flow field (except at possible singularity points). This allows the use of a scalar function, ϕ, to describe the flow field through the definition:

$$\frac{\partial \phi}{\partial x_i} = u_i \tag{4.1}$$

This equation defines each component of the velocity in terms of the local spatial partial derivative in the direction of the velocity component. As stated in Chapter 2 this definition when inserted for velocity in the definition of the vorticity results in the identity that the vorticity is zero, hence irrotational flow.

Continuity Equation

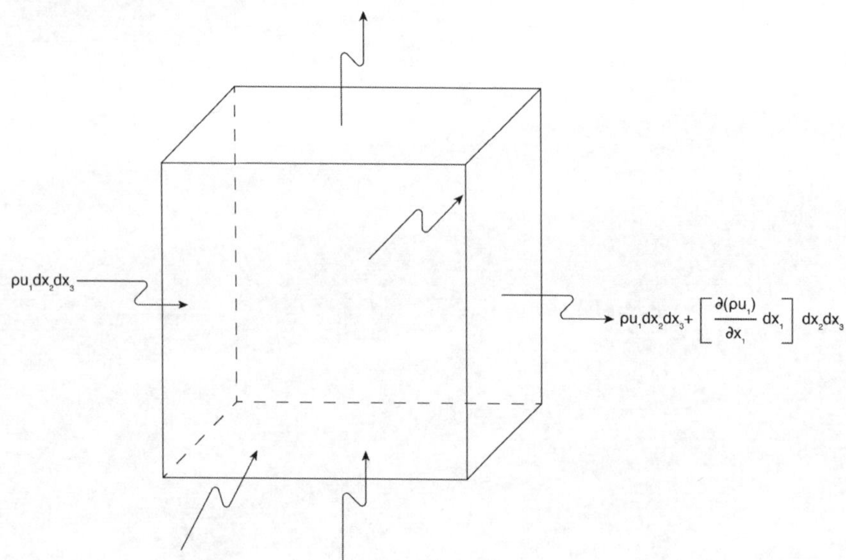

Fig. 4.1 Illustration of a cubic element within a flow field with inflow and outflow at each face; the x_1 direction inflow (right) and outflow (left) is indicated where the outflow is expressed in terms of the change of density times velocity between the outflow and inflow.

Before we get into describing flows with the velocity potential we introduce the continuity equation. This equation comes from conservation of mass as applied to a continuum of fluid that may be in motion. The basic derivation of the continuity equation is shown in Fig. (4.1). Imagine a three dimensional volume in space that for convenience is shaped as a cube. Each face can have mass flow across this geometric element. We are interesting in finding the constraints on the flow field that satisfies conservation of mass for flow in/out of this volume. The basic physics of the relationship that we start from is that no mass can be created or destroyed (thus is conserved) over time. So the net flow in minus the net flow out must be equal to the net change in mass within the volume.

Putting this idea in equation form where the mass flow rate across any specified area for a continuum is the density times the velocity normal to the area times the area. We apply this to the faces of the cube:

Net flow in x_1 (outflow minus inflow):

$$\dot{m}_{out,x_1} - \dot{m}_{in,x_1} = \left[\rho u_1 dx_2 dx_3 + \left(\frac{\partial}{\partial x_1}(\rho u_1)\, dx_1\right) dx_2 dx_3\right] - [\rho u_1 dx_2 dx_3] = \left(\frac{\partial}{\partial x_1}(\rho u_1)\, dx_1\right) dx_2 dx_3 \tag{4.2}$$

This is repeated for the x_2 and x_3 direction where changes of density times velocity are with respect to ∂x_2 and ∂x_3 respectively, and u_2 and u_3 are used for the velocity, respectively, while using the area $dx_1 x_3$ and $d_1 x dx_2$, respectively. Summing all three of these net flow rates results in a scalar representation of the difference between the outflow and inflow where all possible flow paths in and out are included. Reversing the sign (to make it inflow minus outflow) this must equal to the change of mass within the volume element, $dx_1 dx_2 dx_3$.

This is expressed as:

$$\frac{\partial(\rho)}{\partial t} dx_1 dx_2 dx_3 = -\frac{\partial}{\partial x_1}(\rho u_1)\, dx_1 dx_2 dx_3 - \frac{\partial}{\partial x_2}(\rho u_2)\, dx_1 dx_2 dx_3 - \frac{\partial}{\partial x_3}(\rho u_3)\, dx_1 dx_2 dx_3$$

or rearranging and dividing each term by $dx_1 dx_2 dx_3$:

$$\frac{\partial(\rho)}{\partial t} + \frac{\partial}{\partial x_1}(\rho u_1) + \frac{\partial}{\partial x_2}(\rho u_2) + \frac{\partial}{\partial x_3}(\rho u_3) = 0 \tag{4.3}$$

This is the continuity equation that must be satisfied to conserve mass. Notice that if the flow is steady the first term is zero. Also if the density is constant (incompressible) then the first term, or the partial derivative with respect to time, is zero, and density can be factored from each of the other terms and divided out of the equation. The result is:

$$\frac{\partial u_1}{\partial x_1} + \frac{\partial u_2}{\partial x_2} + \frac{\partial u_3}{\partial x_3} = 0 = \frac{\partial u_i}{\partial x_i} \tag{4.4}$$

where in the last term there is summation by tensor notation. This is the reduced form of continuity for incompressible flow. Notice that this form does not require the flow be steady (even though the unsteady derivative of density is no longer included). The velocity may in fact vary with time.

If we now insert the definition of the velocity potential from Eqn. (4.1) for each of the velocity components in Eqn. (4.4) we end up with the following equation for ϕ for incompressible flow:

$$\frac{\partial^2 \phi}{\partial {x_1}^2} + \frac{\partial^2 \phi}{\partial {x_2}^2} + \frac{\partial^2 \phi}{\partial {x_3}^2} = 0 \tag{4.5}$$

This equation is the Laplace operation on the scalar velocity potential, ϕ, and represents continuity (or conservation of mass) for an incompressible flow.

Before moving on we write the continuity equation using the Material Derivative from Chapter 2. We combine the time derivative of density with the other three terms but notice that there is a difference in the three spatial derivative terms from those found in the Material Derivative, the velocity is included in the derivative in continuity. So if each of these terms is expanded,

$$\frac{\partial \left(\rho u_i\right)}{\partial x_i} = \rho \frac{\partial \left(u_i\right)}{\partial x_i} + u_i \frac{\partial \left(\rho\right)}{\partial x_i}$$

The we see that:

$$\frac{D\rho}{Dt} + \rho \frac{\partial \left(u_i\right)}{\partial x_i} = 0$$

And if the density is y constant we obtain Eqn. (4.4) as expected.

Streamfunction

We now introduce the streamfunction, ψ. This is a scalar quantity as is the velocity potential. For simplicity we will do this in two dimensions, but it is valid in three dimensions as well. Recall we defined a streamline coordinate system (s, n) where s is aligned with the velocity vector. In general, for a time dependent flow the streamlines will be continually changing instant by instant. Within a coordinate system (say Cartesian or cylindrical or spherical) the streamline has an equation that can be written out in the selected coordinate system. The equation of this line can be represented by a streamfunction value. This is done as follows.

Consider a line that represents the instantaneous streamline within a flow. In Cartesian coordinates we can write the following for this line where we assume that there is some constant value ψ associated with the equation for the line:

$$\psi = f(x_1, x_2) = \text{constant}$$

For ψ being a constant small changes along a streamline we can write:

$$d\,\psi = 0 = \frac{\partial\psi}{\partial x_1}dx_1 + \frac{\partial\psi}{\partial x_2}dx_2$$

$$\frac{dx_2}{dx_1} = -\frac{\frac{\partial\psi}{\partial x_1}}{\frac{\partial\psi}{\partial x_2}}$$

Since $\frac{dx_2}{dx_1}$ is the slope of the line representing the streamline and since the velocity vector is tangent to the streamline, and the slope of the velocity vector is the ratio of the x_2 to x_1 velocity components we write:

$$\frac{u_2}{u_1} = -\frac{\frac{\partial\psi}{\partial x_1}}{\frac{\partial\psi}{\partial x_2}}$$

or we can write:

$$u_1 = \frac{\partial\psi}{\partial x_2} \text{ and } u_2 = -\frac{\partial\psi}{\partial x_1} \tag{4.6}$$

Eqn. (4.6) represents the definition of the scalar $\psi(x_1, x_2)$ associated with each streamline. If we know expressions for the velocity components as a function of position then we can integrate Eqn. (4.6) to find the value of the streamfunction, ψ. Similarly if we know the equation for the streamfunction then we can calculate the values of each velocity component through partial differentiation using Eqn. (4.6).

Let's assume an incompressible flow so that the flow field follows the continuity equation given by Eqn. (4.4). Now insert for the derivatives of the velocities in terms of the derivatives of the streamfunction. The results is:

$$\frac{\partial^2 \psi}{\partial x_1 \partial x_2} + \frac{\partial^2 \psi}{\partial x_1 \partial x_2} = 0$$

This is an identity, in other words it is automatically true, so the existence of the streamfunction, by the given definition, automatically solves the continuity equation. Said another way, if the streamfunction exists by its definition of Eqn. (4.6) then the flow satisfies continuity for incompressible conditions.

It is possible to take a given velocity field and construct a number of streamlines. At any given point there is a velocity vector and therefor a streamline that passes through it. The only time there can be two or more streamlines passing through a given point (intersecting at some random angle) is if the magnitude of the velocity is zero. Then both partial derivatives of Eqn. (4.6) are zero and the slope is not defined. A stagnation point is such an intersection of streamlines, as shown in Fig. (4.2) for flow over a cylinder. The streamfunction can be continuous up to the stagnation point and beyond, say following the cylinder surface, but it divides at the stagnation point one branch going up and another going down. Stagnation points don't have to be on surfaces they can be distributed within the flow field.

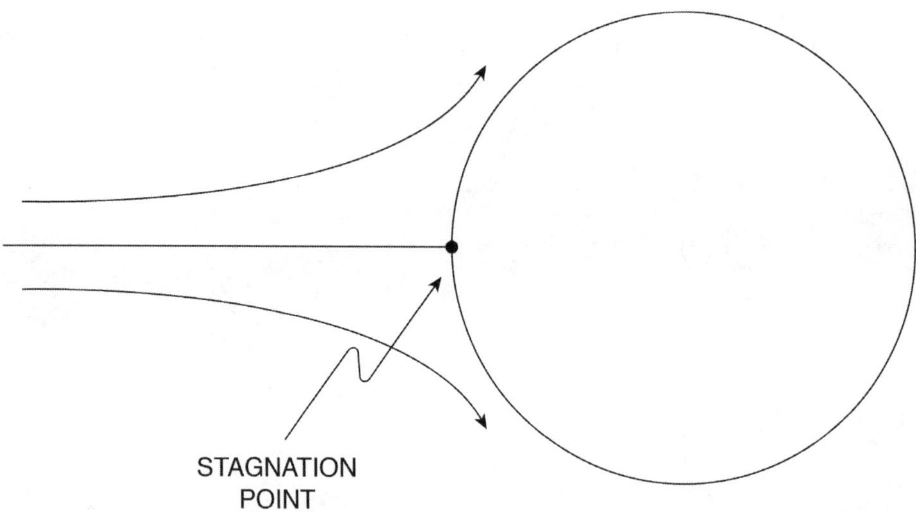

Fig. 4.2 Streamlines illustrating a stagnation point streamline where flow separates away from this point, and the velocity at this point is zero.

Streamfunctions are valuable in that they can provide information on local flow rate conditions within a flow field. In general the flow rate (mass or volume) is determined by the velocity vector and an area through which the flow occurs. That is to say, the velocity vector only provides flow rate through an area if there is velocity vector component normal to the area. For a given area we define an outward normal unit vector, $\hat{n}$ as shown in Fig. (4.3). The mass flow rate through the area "A" with this outward normal is given as:

$$\dot{m} = \rho \boldsymbol{V} \bullet \hat{n} A \tag{4.7}$$

The reader should check the units for this equation. Notice that $\boldsymbol{V} \cdot \hat{n}$ is the dot product between the velocity and outward normal that results in a scalar whose value represents the projection of the velocity vector in the $\hat{n}$ direction. To obtain the volume flow rate, $\dot{Q}$ this expression is divided by mass per volume, or the density:

$$\dot{Q} = \boldsymbol{V} \bullet \hat{n} A \tag{4.8}$$

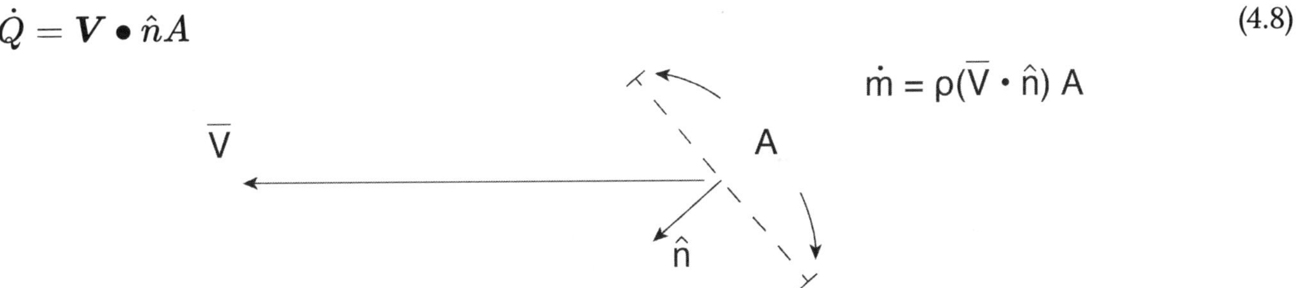

Fig. 4.3 Flow through area A with outward normal $\hat{n}$ and velocity vector $\bar{V}$.

Now consider a two dimensional steady flow with a streamline distribution as shown in Fig (4.4). Since the velocity vector is tangent to each streamline there can be no flow across a streamline. Consequently, the flow that occurs between two streamlines must remain between those two streamlines along the flow direction. In other words, the flow rate between two streamlines remains constant. The value of the flow rate can be interpreted in terms of the change in the streamfunction value between the two streamlines. This is shown as follows.

Consider the two streamfunctions in Fig. (4.4), such that the difference is $\Delta\psi = \psi_2 - \psi_1$. Next draw, a control volume as shown in the figure, where flow can enter through two areas, dx_1 and dx_2 (this is two dimensional representation so there is a unit distance into the page). The volumetric flow rate per unit depth into the control volume must balance the volumetric flow rate per unit depth out of the control volume, $\dot{Q}'$.

$$\dot{Q}' = \int_B^C u_2 dx_1 + \int_C^A u_1 dx_2 = -u_2 \triangle x_1 + u_1 \triangle x_2$$

$$\dot{Q}' = \triangle \psi_{C-B} + \triangle \psi_{A-C} = \triangle \psi_{A-B} = \psi_2 - \psi_1 \tag{4.9}$$

The reason there is a negative sign for $u_2 \Delta x_1$ is that in determining the flow rate between points B and C we integrate along negative Δx_1 direction (or the change in Δx_1 is negative).

Also we have used the definition of the streamfuction, Eqn. (4.7) to evaluate finite changes, on $u_1 = \triangle\, \psi / \triangle\, x_2$ and $u_2 = -\triangle\, \psi / \triangle\, x_1$. The interpretation then is that the flow rate $\dot{Q}'$ is equivalent to the change in streamfunction value between two points within the flow.

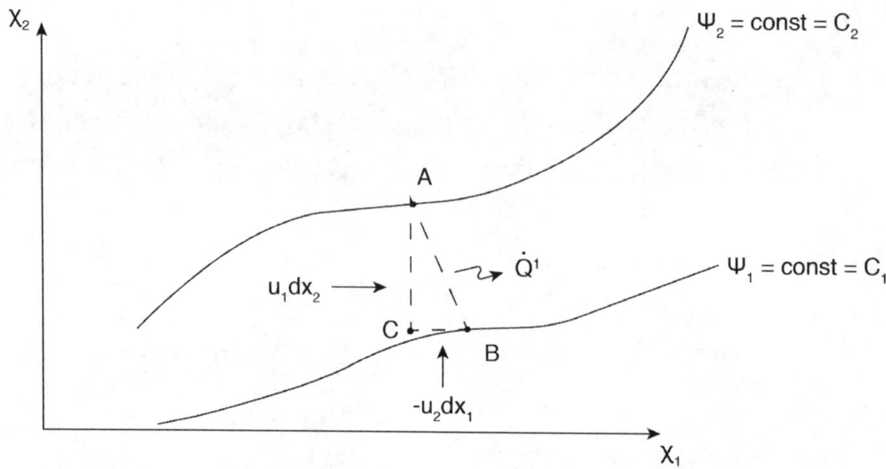

Fig. 4.4 Flow between two different streamlines illustrating constant flow rate between the streamlines; control volume ABC is used to show relationships between flow rate and change in streamfunction values, Eqn. 4.9.

Interestingly, one can use streamline maps to qualitatively and quantitatively evaluate the velocity field. Imagine a wind tunnel test in a two dimensional flow over, say, a wing, as in Fig. (4.5). Smoke dye is injected at discrete points upstream separated by some vertical distance between each streamline. The lines of smoke travel downstream and over the wing. As the flow goes over the wing some of the streamlines diverge and some converge (the distance of separation between streamlines changes). Since we have shown that there is constant flow rate between streamlines when the distance between streamlines gets smaller the area of the flow decreases, so the velocity must increase. As streamlines diverge the velocity must decrease. The relationship between cross sectional area and velocity is linear as shown in Eqn. (4.8). Measuring the change of distance between adjacent streamlines provides a measure of the amount of increase or decrease of velocity.

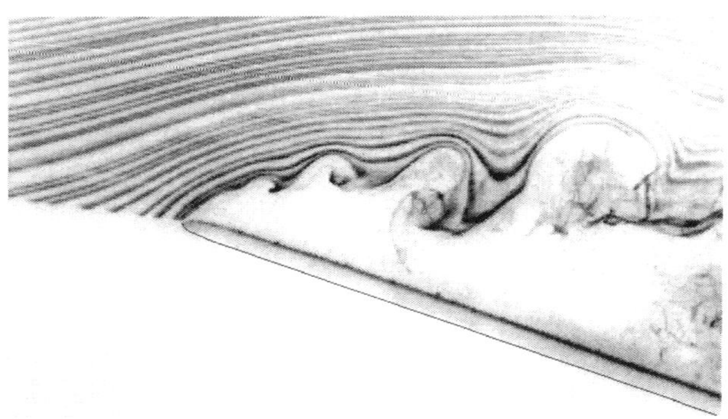

Fig. 4.5 Smoke flow visualization of flow over an inclined flat wing.

We now have a physical as well as mathematical interpretation for the streamfunction. Remember that this is a scalar field function representative of the local velocity. If we use the definition of streamfunction, Eqn. (4.7) and insert this into the definition of vorticity for a two dimensional flow in the $x_1 - x_2$ plane we obtain the following:

$$\omega_3 = \left(\frac{\partial u_2}{\partial x_1} - \frac{\partial u_1}{\partial x_2}\right) = \left(-\frac{\partial^2 \psi}{\partial {x_1}^2} - \frac{\partial^2 \psi}{\partial {x_2}^2}\right) = -\left(\frac{\partial^2 \psi}{\partial {x_1}^2} + \frac{\partial^2 \psi}{\partial {x_2}^2}\right) \tag{4.10}$$

We see that the vorticity is equal to the negative of the Laplace of the streamfunction (Shown here in two dimensional flow, but is also the case in three dimensional flow). For $irrotational flow$ the vorticity is identically zero so:

$$\left(\frac{\partial^2 \psi}{\partial {x_1}^2} + \frac{\partial^2 \psi}{\partial {x_2}^2}\right) = 0$$

This shows that the Laplace of the streamfunction is zero for irrotational flow and follows the same results for the velocity potential for incompressible flow. So for irrotational, incompressible (ideal) flow the Laplace of both the velocity potential and streamfunction are equal to zero. This points to the ability to solve the Laplace equation for either of these quantities and from this solution determine the velocity field from the definitions of ψ and ϕ in teams of the velocity. This will be the approach we take in the next Chapter.

Since both ϕ and ψ have the same functional form one might think that they are related to each other. We see this in comparing Eqns. (4.1) and (4.7), both are related to velocity derivatives. Notice that

$$u_1 = \frac{\partial \phi}{\partial x_1} = \frac{\partial \psi}{\partial x_2} \text{ and } u_2 = \frac{\partial \phi}{\partial x_2} = -\frac{\partial \psi}{\partial x_1}$$

The velocity is tangent to the constant streamfunction value, but the velocity is normal to the constant velocity potential value. Consequently lines of constant ϕ are normal to lines of constant ψ. It is straight-forward to show that the slope of constant potential lines is $-\frac{u_1}{u_2}$ while the slope of constant streamfunction lines is $\frac{u_2}{u_1}$. We will generate plots for specific flows in the next chapter to illustrate this, but shown in Fig. 4.6 is an illustration of the orthogonality of the velocity potential lines relative to the streamfunction lines. Velocity is always along the streamfunction line and normal to the potential lines.

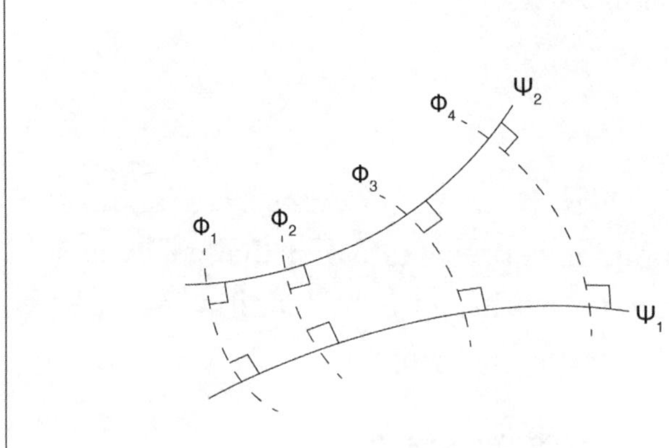

Fig. 4.6 Illustration of streamlines (lines of constant ψ) and velocity potential lines, ϕ, indicating the orthogonal condition of the two lines.

In Table 4.1 are the two dimensional expressions in cylindrical coordinates for the various mathematical representations presented here. Note that v_r is the radial velocity component and v_r is the circumferential velocity component. The vorticity only has a z component, as shown, all others are identically zero for a two dimensional flow in r, Θ.

Table 4.1 Cylindrical Coordinate Representation for Incompressible Flow

Continuity: $\frac{\partial (rv_r)}{r\partial r} + \frac{\partial v_\theta}{r\partial \theta} = 0$

Streamfunction: $v_r = \frac{\partial \psi}{r\partial \theta} \qquad v_\theta = \frac{\partial \psi}{\partial r}$

Vorticity: $\omega_z = \left(\frac{\partial\left(rv_\theta\right)}{r\partial r} - \frac{\partial v_r}{r\partial\theta}\right)$

Velocity Potential: $v_r = \frac{\partial\phi}{\partial r} v_\theta = \frac{\partial\phi}{r\partial\theta}$

V. Potential Flows

In this chapter we will introduce a number of "basic ideal flows". These flows will form a basis from which we can construct more complex flows. The basic principle we are relying on is "superposition". This allows the linear addition of various flows that then result in more complicated flows. This is possible because the basic underlying equations that govern the flows are linear.

We have shown that there is an orthogonal relationship that exists between two variables that describe the flow, streamfunction and velocity potential, for ideal (irrotational and incompressible with zero viscous forces) flows. The orthogonal condition indicates that if we know one of these it is rather straight-forward to determine the other. We will use both of these flow descriptors to some extend, but mostly use the streamfunction representation of the various flows that we will consider. We will also restrict our results to two dimensional cases, although this is not necessary in general. Our basic equations are the Laplace equations we found in the previous chapter for the streamfunction, ψ, and velocity potential, ϕ.

For ideal flows we have the simplified continuity equation that treats the density as a constant, and allows the elimination of the density directly in the equation. This results in a relationship among the velocity components that must hold true to satisfy conservation of mass. Also, the irrotational flow condition requires the vorticity to be zero which leads to additional conditions on velocity derivatives within the flow. The boundary conditions need to be specified for such flows, and are required to solve the governing equations. Since these flows are inviscid we do not have the no-slip boundary condition to help specify the value of the velocity. This implies that the velocity may be some (finite) value at a surface, not equal to the surface velocity). However, we can require that a surface be impermeable, that is no flow crosses the boundary. This then assures that the component of the velocity normal to the surface must be zero. So at least we can say something quantitative about a velocity component.

Basic Flows

In this section we present the governing equations for several basic flows. These equations are solutions of the Laplace equation and are determined through required boundary or imposed flow conditions. We deal with steady two dimensional flows.

Uniform Flow

The most simple flow (other than zero flow) is a steady uniform flow. This condition is a constant velocity in a given direction such that the velocity vector does not vary spatially. We designate this velocity as U. If we align our coordinate system along the direction of U, such that x_1 is the direction of U, then there is only one nonzero velocity component. The streamfunction would be expected to be a straight line along the direction of U as well. Using the definition of ψ we obtain the following:

$$u = U = \frac{\partial \psi}{\partial x_2}$$

Integrating this in x_2 we obtain an expression for the streamfunction:

$$\psi = U x_2 + f(x_1)$$

But since

$$v = 0 = -\frac{\partial \psi}{\partial x_1}$$

then ψ is not a function of x_1 and we obtain:

$$\psi = U x_2 + C \qquad (5.1)$$

We can arbitrarily set $\psi = 0$ at $x_2 = 0$ so that

$$\psi = U x_2 \qquad (5.2)$$

If we proceed along the same line to solve for ϕ from its definition relative to the partial velocity derivative, Eqn. (4.1), the result is:

$$\phi = U x_1 \qquad (5.3)$$

where we have set $\phi = 0$ at $x_1 = 0$, other wise there is an additive constant which is equal to the potential at $x_1 = 0$.

For a condition of uniform flow, again with velocity of U, but at some angle α to the x_1 axis we have the following (the reader is encouraged to obtain this result by integrating the definition of the streamfunction):

$$\psi = U\left(x_2 cos\alpha - x_1 sin\alpha\right) + C$$

(5.4)

The constant C is determined by a "datum" value of ψ that passes through some point. For instance if $\psi = 0$ at $(0, 0)$ then $C = 0$.

Going back to uniform flow (two dimensional) aligned in the x_1 direction we can find the volumetric flow rate per depth through some area say between $x_2 = 0$ and $x_2 = 4$ as:

$\dot{Q}' = \Delta\psi = U(4 - 0) = 4U$

Source/Sink Flow

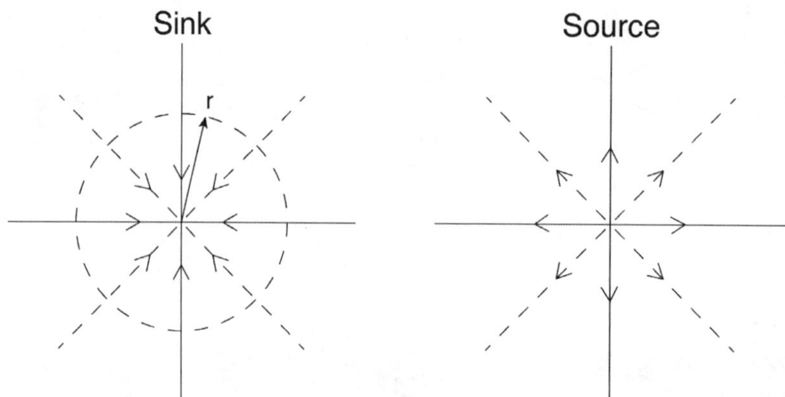

Fig. 5.1 Sink and source flows from a point, a sink is the negative of the source and has an inward radial velocity.

In two dimensional flow a source or a sink of flow is possible, since it implies that flow enters or leaves a given two dimensional plane. We treat this as a rate of gain (source) or loss (sink) of mass at some point within the flow. An example of this might be a simulation of a drain say in a tank with a flat bottom. Along the bottom plane flow leaves the plane of the bottom at a small opening. Here we shrink the size of the opening to a point (which is hard to imagine since we have mass flow rate through this point, since the velocity associated with this point is infinite across the plane). Despite these unrealistic characteristics, if we move away from this "singularity" point of infinite velocity and zero area we can assign a flow rate. This is illustrated in Fig. 5.1, which illustrates how flow approaches the sink along the plane with velocity vectors that are inwardly, straight radial streamlines, as shown as vector lines in Fig. 5.1.

By drawing a circle of an arbitrary radius, r, shown as the dashed line in Fig. 5.1, the flow rate is determined by integrating the velocity around the circle. This is because the circle represents the flow area (per unit depth into the page), and the velocity vectors are each normal to the circle based on symmetry. The result for the mass flow rate per unit depth into the page is:

$$\dot{m}' = \rho v_r 2\pi r$$

Note that v_r is negative for a sink and positive for a source.

If a larger circle is drawn the same mass flow rate occurs into the point, but since the area is larger the velocity at the new larger circle is less. One can solve for the velocity, v_r, at any radial position, r, as:

$$v_r = \frac{\dot{m}'}{2\pi\rho}\frac{1}{r} = \frac{\dot{Q}'}{2\pi}\frac{1}{r} \tag{5.5}$$

For a constant steady volumetric flow, $\dot{Q}'$, we see that the velocity decreases linearly as r increases. So this flow has straight streamlines all represented as radial lines from the center point. We see how the velocity increases rapidly in magnitude as r goes to zero, at which point it becomes infinite (or undefined). Based on the orthogonal condition between the velocity potential and streamfunction we see that lines of constant potential are circles (like the dashed line shown in Fig. 5.1.)

We can combine the constants in Eqn. (5.5) $\mu_s = \frac{\dot{Q}'}{2\pi}$, where μ_s is the "source strength" with units of m^2/s in SI units such that:

$$v_{r,source} = \frac{\mu_s}{r} \tag{5.6}$$

Eqn. (5.6) represents the flow from a source with strength μ_s. The source pumps fluid into the plane of flow with streamlines that are radially outward.

The flow can be reversed such that the flow along the radial lines is inward. This implies flow is exiting the plane, and this is a "sink", with velocity inward, which has a negative $r$component:

$$v_{r,sink} = -\frac{\mu_s}{r} \tag{5.7}$$

The streamfunction is found by integrating the velocity based on the definition of the streamfunction using cylindrical coordinates:

$$v_r = \frac{\partial\psi}{r\partial\theta} = \frac{\mu_r}{r}$$

$$\psi = \mu_s \theta \tag{5.8}$$

Velocity potential lines are found from the definition of the potential relative to the velocity:

$$v_r = \frac{\partial \phi}{\partial r} = \frac{\mu_s}{r}$$

Integrating this results in:

$$\phi = \mu_s ln\, r \tag{5.9}$$

Vortex Flow

Now consider a swirling flow such that the streamlines are circles. This implies that there is no radial velocity component, only v_θ. We label this swirling flow as a "vortex", often called a "free vortex" since it is free from external forcing. It is also irrotational. If streamlines are circles as is shown in Fig. 5.2 then velocity potential lines must be straight radial lines from the center of the circle to be orthogonal to the streamlines.

To construct a flow with the above characteristics we examine the possible flow in cylindrical coordinates. Here we use the definition of the streamfunction in cylindrical coordinates:

$$v_\theta = -\frac{\partial \psi}{\partial r}$$

$$\omega_z = 0 = \frac{\partial \left(r v_\theta \right)}{\partial r}$$

So if we consider the flow to be given by its velocity $v_\theta = c/r$, where

C is a constant, then the vorticity will be zero as required. Inserting this into the definition for the streamfunction above yields:

$$\psi = -C ln\, r \tag{5.10}$$

Next we define the "circulation" Γ, as the line integral of the velocity around a closed line, or a loop, as:

$$\Gamma = \oint v_\theta ds \tag{5.11}$$

This has units of m^2/s in SI units. Inserting the expression for v_θ above and noting that $ds = rd\theta$ with the integration carried out between 0 and 2π, we get:

$$\Gamma = \oint v_\theta ds = 2\pi C \tag{5.12}$$

The result is that the circulation within the vortex is a constant that can be determined knowing C. We can replace C with rv_θ and solve for the velocity as:

$$v_\theta = \frac{\Gamma}{2\pi r} = \frac{\mu_v}{r} \tag{5.13}$$

where we define the "vortex strength" as:

$$\mu_v = \frac{\Gamma}{2\pi} \tag{5.14}$$

And also using the definition of vorticity, which must be zero to be irrotational, as:

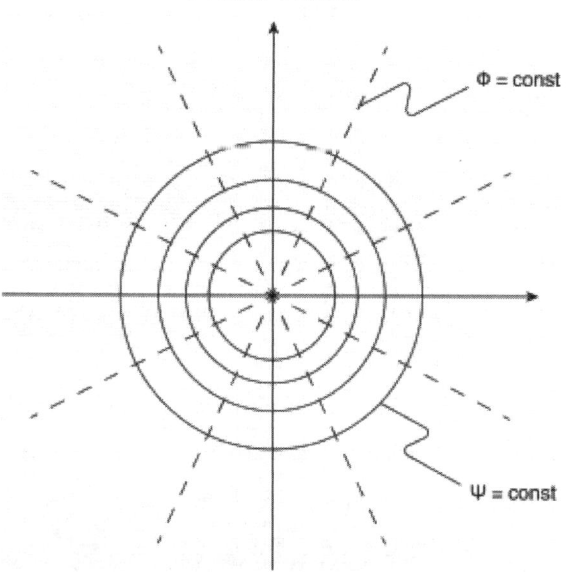

Fig. 5.2 Sketch of the streamlines and potential lines for a free vortex; note that the circumferential velocity decreases in the radial direction.

The flow for a free vortex is shown in Fig. 5.2 indicating streamlines as circles and the velocity potential as straight radial lines. Notice that theses lines are just the inverse of what we found for the source.

Since streamlines can have the velocity direction either counterclockwise or clockwise and still have the same general form with the same equations as shown above the sign convention is that a counterclockwise rotation is positive, and a clockwise rotation is negative. This is expressed in the value of Γ, or μ_s. In other words counterclockwise flow has positive circulation.

One last interesting aspect of the free vortex and the circulation is as follows. We use the vector identity:

$$\int (\nabla \times \boldsymbol{V})\, dA = \oint \boldsymbol{V} \cdot \boldsymbol{ds}$$

where $\boldsymbol{V}$ is any vector, dA is an area and $\mathbf{ds}$ is the vector distance along a closed loop integration around the area A. If we let $\boldsymbol{V}$ be the velocity and noting that the vorticity is $\boldsymbol{\omega} = (\nabla \times \boldsymbol{V})$ and that the loop line integral is the circulation given above, then we can see that:

$$\Gamma = \oint v_\theta ds = \int \boldsymbol{\omega} dA \tag{5.15}$$

Consequently the circulation is the area integral of the vorticity in some selected area within the flow. Note that we are dealing with a two dimensional flow and dA is within the plane of the flow so the vorticity is aligned along a vector out of the plane.

Now we have a bit of a dilemma. A free vortex has finite values of velocity v_θ that result is a certain circulation. This circulation is proportional to the vorticity within the flow. But we have said that the flow is irrotational, which means that the vorticity is zero. How can the circulation be nonzero while the vorticity is zero? The answer is that the vorticity (that drives the circular velocity) is concentrated at the center of the circle. Away from the center of the circle we know the vorticity is zero since the velocity we are using, v_θ, was based on the vorticity being zero. If one calculates the circulation about a circular loop of radius r_1 (an area of $\pi {r_1}^2$) and then repeats this calculation for a larger area of radius r_2, the result will be the same. This shows that in the area between r_1 and r_2 there is no added vorticity since the circulation remains the same. This can be done for any arbitrarily small radius r_1, showing that in the limit of r going to zero there is no vorticity within the flow except at r=0. A free vortex has a concentration of vorticity at r = 0. Also, the velocity must go to infinity, a velocity singularity. Interestingly this says that we can have isolated vorticity concentrations within an otherwise irrotational flow field.

Superposition

As stated previously superposition is a powerful tool that allows us to construct more complex flows from several simple flows. This is possible since the governing equation for the streamfunction, that describes streamlines, and the velocity potential are linear, is the Laplace equation.

We begin by examining the flow field in the vicinity of a source and sink. We place the source and sink on the x_1 axis separated by distance 2a, as is shown in Fig 5.3 (a), with the origin mid way between each. The source is on the left (negative x_1, and the sink is on the right (positive x_1). The streamfunction,or the velocity potential, at any point in the flow can be obtained by added together the streamfunction, or velocity potential, from the source plus that for the sink. However, care must be taken into account since our previously obtained equations were written assuming the source/sink were located at the origin of our coordinate system. In this case they are shifted along the x_1 axis. The result for any point P, located at (x_1, x_2) in the flow at distance r_1 and r_2 from the sink and source, respectively is:

$$\psi = \psi_{source} + \psi_{sink}$$

$$\psi = \mu_s \theta_1 - \mu_s \theta_2 \tag{5.16}$$

where it is assumed the source and sink have equal but opposite strength and the angles are shown in Fig. 5.3. Similarly for the velocity potential:

$$\phi = \mu_s ln\, r_2 - \mu_s ln\, r_1 = \mu_s ln\, \frac{r_2}{r_1} \tag{5.17}$$

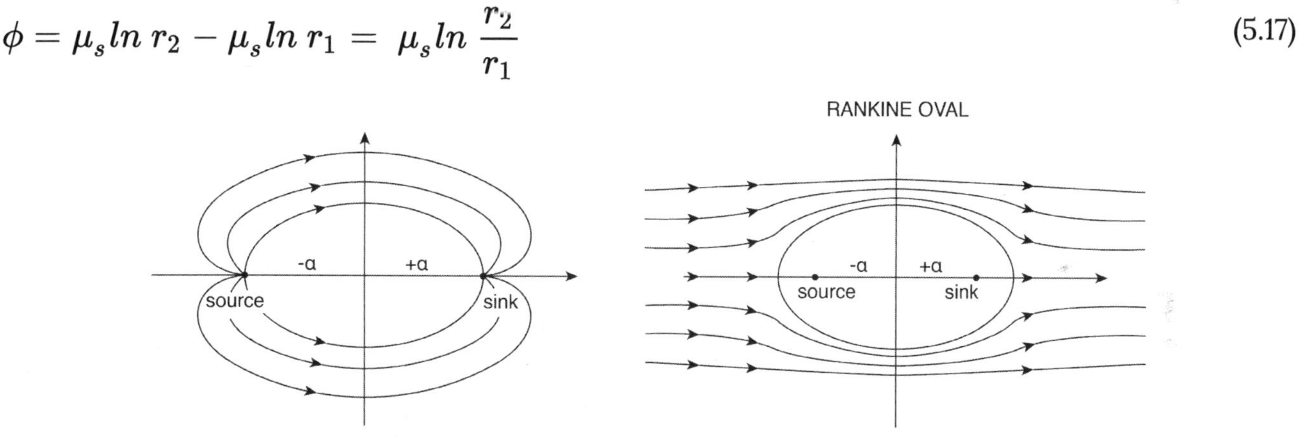

Fig. 5.3 (a) Superposition of a source and sink of equal strength both positioned on the x_1 axis distance "a" from the origin, the source on the left and sink is on the right; (b) superposition of a uniform flow, a source and a sink which creates a Rankine Oval

In order to arrive at an equation for the streamfunction for the combined flow we use a trig identity:

$$tan(\theta_1 \pm \theta_2) = \frac{tan\theta_1 \pm \; tan\;\theta_2}{1 \mp tan\;\theta_1 tan\;\theta_2}$$

So taking the arctan of both sides and noting that $tan\;\theta_1 = \frac{x_2}{x_1 + a}$ and $tan\;\theta_2 = \frac{x_2}{x_1 - a}$, so that the right hand side of the above equation becomes: $arctan\left(\frac{2ax_2}{{x_1}^2 - a^2 + {x_2}^2}\right)$, the streamfunction is:

$$\psi = -\mu_s \; arctan\left(\frac{2ax_2}{{x_1}^2 - a^2 + {x_2}^2}\right) \tag{5.18}$$

For example, the streamfunction at $P = (3, 4)$ for {a} = 2 is $\psi = -\mu_s(37.3^o)$.

At this point we have a source and sink separated by a distance of 2a. Extending this we can add a uniform flow in the positive x_1 direction to the combined source sink flow to obtain the following:

$$\psi = \; Ux_2 - \mu_s \; arctan\left(\frac{2ax_2}{{x_1}^2 - a^2 + {x_2}^2}\right) = Ur\;sin\;\theta + \mu_s\left(\theta_1 - \theta_2\right) \tag{5.19}$$

where the last term is the result written in cylindrical coordinates. If we set the streamfunction equal to some constant we can plot the associated streamline from Eqn. (5.19). In particular setting $\psi = 0$ an oval results as shown in Fig. 5.3 (b). This is known as the *Rankine Oval*. The characteristics of this oval can be adjusted by inserting different values for μ_s for a given U. The characteristics of the oval are:

Major axis $(x_2 = 0) = \; 2a\left(1 + \frac{2\mu_s}{a}\right)^{1/2}$

Minor axis $(x_1 = 0) = 2a\left(cot\;\frac{h/2a}{2\mu_s/aU}\right)$ (where h= minor axis) (5.20)

It is possible then to model the flow over an oval surface with an approach velocity of U by Eqn. (5.19). The velocity is determined by taking the derivatives of ψ relative to x_1, and x_2. The geometry of the oval can be adjusted by varying the strength of the source/sink as well as their locations, a.

DOUBLET:

A doublet is a result of construction of a flow field using the superposition of a source and a sink that are placed very close to each other. The superposition of these two will result in flow leaving the source and entering the sink. The flow lines form circular paths as the flow attempts to leave the source while being drawn in to the sink. By making the strength of the source and sink identical a symmetric flow will result. It will be shown how this flow establishes a streamline that is a circle in the limit of the source and sink approaching each other spatially. This type of flow has powerful applications to simulate more complicated flows as we will see.

To extend the use of a source and sink of equal strength to a doublet we take the limit as their separation distance $a \to 0$. First we note that the arc tan of a small number is equal to the number so $arctan\left(\frac{2ax_2}{{x_1}^2 - a^2 + {x_2}^2}\right) = \frac{2ax_2}{{x_1}^2 - a^2 + {x_2}^2}$ then we have:

$$\psi = -\frac{2a\mu_s x_2}{{x_1}^2 - a^2 + {x_2}^2}$$

Since we want $a \to 0$ we will also let $\mu_s \to \infty$ at the same time and say $a\mu_s = C$, where C is some constant. This may seem arbitrary but it assures us that the streamfunction doesn't vanish to zero and since we can set μ_s to any desired value. The result is that we can rewrite the equation for the streamfunction of a doublet as:

$$\psi = -\mu_d \frac{x_2}{{x_1}^2 + {x_2}^2} \tag{5.21}$$

where $\mu_d = 2C$ a constant that determines the "strength of the doublet". This equation can be rearranged:

$${x_1}^2 + \left(x_2 + \frac{\mu_d}{2\psi}\right)^2 = \left(\frac{\mu_d}{2\psi}\right)^2$$

Noting that for a constant value of ψ the coefficient on the right inside the parenthesis, and the same term added to x_2 is a constant resulting in an equation of a circle. The center of the circle changes with changing values of ψ as does the radius of the circle. The result is a series of streamlines for selected values of ψ, each of which is a circle centered along the x_2 axis as is shown in Fig. 5.4. Notice that the radius of the circle is $\frac{\mu_d}{2\psi}$.

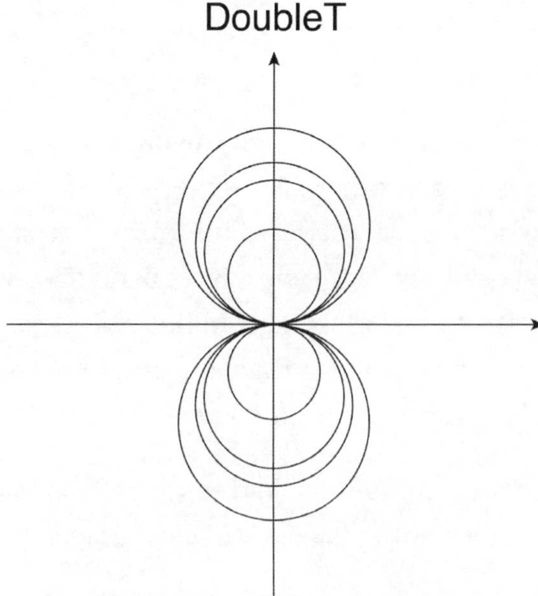

Fig. 5.4 Streamlines associated with a doublet – a source and sink with a distance a approaching zero

Uniform Flow over a Cylinder

The doublet can be added to a uniform flow in the positive x_1 direction resulting in a streamfunction given by:

$$\psi = Ux_2 - \frac{\mu_d x_2}{{x_1}^2 - a^2 + {x_2}^2} \tag{5.22}$$

This equation can be recast in cylindrical coordinates where $r^2 = {x_1}^2 + {x_2}^2$, $x_1 = rcos\theta$ and $x_2 = rsin\theta$ and we define $a = \left(\frac{\mu_d}{U}\right)^{1/2}$, a constant for a given value of U, the result is:

$$\psi = Ur\ \ sin\theta - Ua^2 \frac{sin\ \theta}{r} = Ur\ sin\theta \left(1 - \frac{a^2}{r^2}\right) \tag{5.23}$$

It can be seen that in the limit of large values of r the flow reverts back to uniform flow and the contribution for the doublet goes to zero. If we let $\psi = 0$ then $r = a$, a constant. That is to say, the streamfunction of $\psi = 0$ is a circle of radius a. So we have now constructed the flow field for uniform flow over a cylinder of radius a.

From this the velocity components in the r, θ coordinates are found to be:

$$v_r = \frac{\partial \psi}{r \partial \theta} = U\, cos\theta \left(1 - \frac{a^2}{r^2}\right)$$

$$v_\theta = -\frac{\partial \psi}{\partial r} = -U \sin\theta \left(1 + \frac{a^2}{r^2}\right) \tag{5.24}$$

and the velocity vector is $V^2(r, \theta) = {v_r}^2 + {v_\theta}^2$. The streamline distribution is shown in Fig. 5.5. We are only interested in the flow on the outside of the circle, which represents a cylinder. It is important to note that at $r = a$, the velocity is not zero and depends on θ. So at different positions around the cylinder surface the velocity will change. The largest velocity occurs at $\theta = 90^o$ and 270°, which represents the top and bottom of the cylinder. This is where the streamlines converge the most indicative of high velocity flow. Also at $\theta = 0^o$ and 180° the velocity is zero. These are stagnation points on the cylinder. Notice also that the streamlines are symmetric about the x_1 and x_2 axes. This has important implications on the forces that exist on the cylinder caused by the flow.

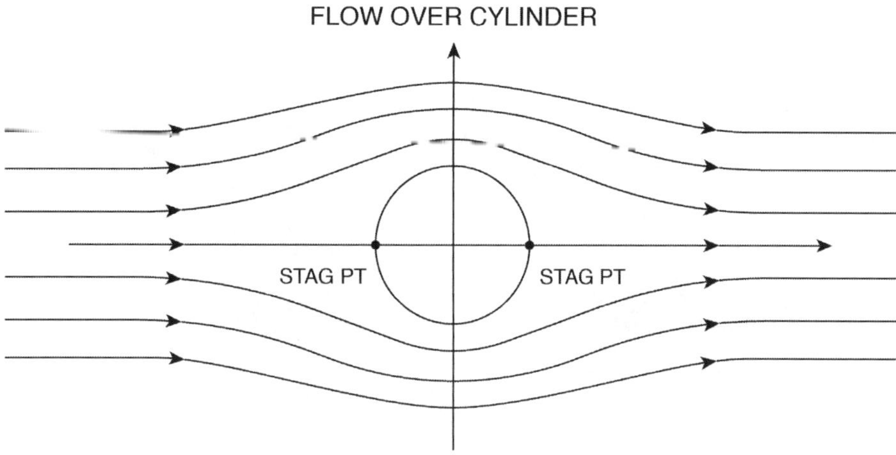

Fig. 5.5 Streamline for flow over a cylinder formed from the superposition of a doublet with uniform flow; note stagnation on either side of the cylinder

IMAGING

There is a method, call imaging, that allows for the insertion of impervious walls or boundaries within a flow. The basic idea here is that a streamline can be used to simulate a solid boundary since it does not allow flow to cross the streamline location. Consequently, if basic flow elements can be combined to yield a streamline that is of the desired shape of a boundary then it satisfies the required flow. For instance, in the condition of uniform flow over a cylinder described above flow elements are combined (uniform flow and a doublet) that result in a simulation of flow over a cylinder, of radius.

Consider the case of a source flow in the vicinity of a flat wall. This may be a simulation of pumping fluid from a well into the surrounding porous region near a solid boundary, such as a rock formation. This could be reversed and have flow into the well (sink). This is shown in Fig. 5.6 where the source is some distance away from the flat wall. We take the wall to be the x_1 axis and the source to be a distance "l": away from the wall. If we make the strength of a "mirror image" of the source on the other side of the wall the flow becomes symmetric about the wall with the wall being a streamline (no flow crosses it). Note here that the velocity component in the x_2 direction at the wall is zero but not the x_1 component. Without providing further details the superposition results in the following equation for the streamfunction:

$$\psi = \mu_s (\theta_1 + \theta_2)$$

θ_1 and θ_2 are measured from the source and its image, respectively, to any point within the flow. This is similar to what was done when combining a source and sink resulting in the Rankine Oval. In fact the analysis following Eqn. (5.16) can be repeated for this flow but the sign in front of θ_2 replaced with the positive sign shown above for ψ.

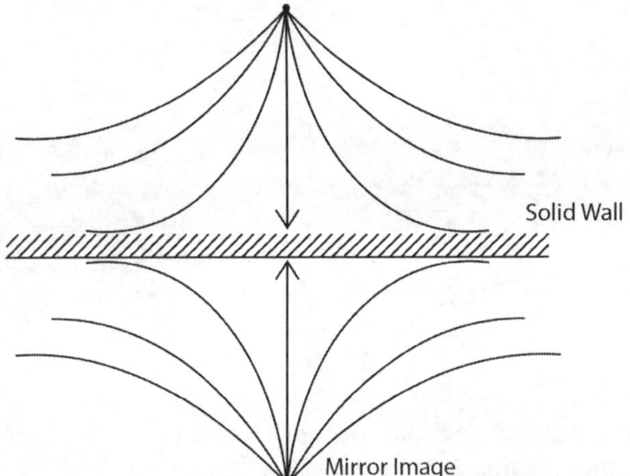

Fig. 5.6 Streamlines of a source near a solid impermeable boundary formed using a mirror image, the wall is a streamline from the superposition of an identical source at an equal distance from the wall on the opposite side.

We introduce one more superposition, that of uniform flow, a doublet and a vortex. The combined streamfunction is:

$$\psi = Ur\ sin\ \theta\ - \mu_d\ \frac{sin\ \theta}{r}\ - \mu_v ln\ r\ + C \tag{5.25}$$

where C is a constant that can be determined by setting the value of ψ at some point in the flow, as shown below. The coordinate system used here has its origin at the center of the circle generated by the uniform flow and doublet. The added vortex does not add a radial component of velocity, since its flow streamlines are all circles. The result is adding a θ component of velocity throughout the flow that depends on the radial location. In Eqn. (5.25) the direction of the added circumferential flow is clockwise (negative). To summarize the strengths of the elements we have $U, \mu_d = Ua^2$ and $\mu_v = \frac{\Gamma}{2\pi}$. We can think of these as adjustable parameters to the flow field.

Rewriting Eqn. (5.25) by combining the first and second term as was done for flow over a cylinder and then setting $r = a$, inserting the expression for μ_d, as well as setting $\psi = 0$, we have:

$$C = \mu_v ln\ a$$

This then sets $\psi = 0$ on the circle with radius a. Inserting this value of C in Eqn. (5.25) we obtain:

$$\psi = U\ sin\ \theta\ \left(r - \frac{a^2}{r}\right) - \mu_v ln\ \frac{r}{a} \tag{5.26}$$

This represents uniform flow over a rotating cylinder as shown in Fig. 5.7, where the streamline representing the cylinder has $\psi = 0$. Notice that since we have a clockwise (negative) circulation the cylinder is rotating clockwise. The velocity on the surface of the cylinder now has a contribution caused by the vortex in addition to the velocity found for a non-rotating cylinder. We see that the stagnation points have moved away from along the x_1 axis. Since we know that the velocity is zero at the stagnation points we can solve for their location (note that $v_r = 0$ everywhere on the cylinder).

$$v_\theta(r = a) = -\frac{\partial\psi}{\partial r} = -2U sin\ \theta\ + \frac{\Gamma}{2\pi a} \text{ or} \tag{5.27}$$

$$sin\theta_{stag} = \frac{\Gamma}{4\pi aV} \text{ for } V_\theta = 0$$

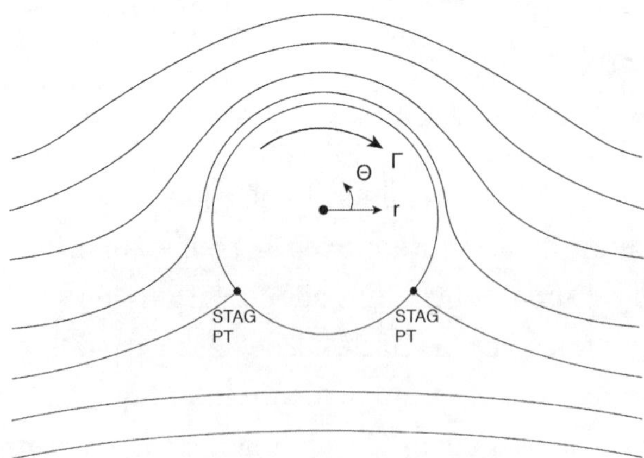

Fig. 5.7 Streamlines for flow over a rotating cylinder formed from uniform flow, a doublet and a vortex; note that the rotation direction determines whether flow accelerates over the top or bottom of the cylinder.

We can also determine the pressure distribution around the cylinder surface using Bernoulli's equation:

$$P(r=a) \ = P_{\infty} + \frac{1}{2}\rho U^2 - \frac{1}{2}\rho {v_\theta}^2 - \rho g h \tag{5.28}$$

where h is the local height on the circle above the centerline (datum)

$h = a \; sin \; \theta.$

We insert Eqn (5.27) for the velocity on the surface, v_θ, into (5.28) to obtain the expression for the local surface pressure. The integral of the pressure around the circle then determines its net force. First, we find the force component in the x_2 direction, this is denoted as the "lift force" per distance into the page, F'_L:

$$F'_L = \ - \oint P(r=a) \ \ sin \; \theta \; a \; d\theta$$

Inserting the expression for P and noting that integrals of odd powers of the sine functions around the entire circle are zero we are left with:

$$F'_L = \ - \rho U \Gamma \ + \ \rho g \pi a^2 \tag{5.29}$$

The last term represents the net body force by the fluid on the volume of the circle (per distance into the page) or the buoyancy force. The lift force is usually defined without the buoyancy force included so we write:

$$F'_L = \; -\rho U \Gamma \tag{5.30}$$

A few things should be stated about this result. First, recall that the viscous force is not included so this is only due to the pressure distribution. Also, when the circulation is zero there is no lift force (there is still a buoyancy force however). We can conclude that the circulation provides the means to create asymmetric conditions around the circle so that a net pressure force occurs. This expression is often called the Kutta-Joukowsky Law who showed this equation to hold for other shapes as well, and is often used in aerodynamics to determine the lift force on two dimensional airfoils based on the circulation associated with the flow around the wing. The sign convention is such that a counterclockwise rotation results in a downward force, and a clockwise rotation results in a upward force for flow along the positive x_1 axis. It is surprisingly accurate for real flows considering the restrictions on its application. This tends to indicate that viscous forces are small at best. It is only accurate for flows that have not separated from the object surface. We will discuss separation when we get into viscous flow affects.

The Kutta-Joukowsky Law (or theorem) as applied to an airfoil requires what is known as the Kutta condition. This is a condition on the flow at the trailing edge that says that the flow exits the airfoil on the top and bottom surfaces of the airfoil with equal velocity and pressure. This implies that the flow does not tend to wrap around the back end or trailing edge of the airfoil and is the boundary condition that allows the calculation of the lift using Eqn. (5.30).

If interested see the web site: https://en.wikipedia.org/wiki/Kutta_condition

Complex Variables for Potential Flow Analysis

In this section the analysis methods presented for potential flow are expanded using some additional mathematical tools that allows for complex representation of flows and illustrates how complex flows can be analyzed. The flow itself is restricted to the conditions associated with potential flow which allows flows to be evaluated using superposition of the Laplace equation.

Basic Formulation of Complex Variables

For ideal flows we focus on the use of the velocity potential and streamfunction, both of which adhere to the Laplace equation, the former representing the conservation of mass, and the latter indicating irrotational flow. Both can be expressed as:

$$\nabla^2 f = 0$$

where f represents either the velocity potential, ϕ, or the streamfunction, ψ. The introduction of either of these variables to define the flow field basically replace the velocity vector as the variable of interest. In arriving at a representation of the flow using complex variable notation the variable z is defined as:

$$z = x + iy$$

or in cylindrical coordinates as:

$$z = r\ (cos\theta + i\ sin\theta) = re^{i\theta}$$

Here, i is the traditional representation of $i = \sqrt{-1}$. Next the Cauchy-Riemann conditions for the two variables ϕ and ψ are introduced, which is predicated on the fact that these two variables satisfy the Laplace eqn. and are thus harmonic functions. The Cauchy-Riemann conditions as:

$$u = \frac{\partial \emptyset}{\partial x} = \frac{\partial \psi}{\partial y}$$

$$v = \frac{\partial \emptyset}{\partial y} = -\frac{\partial \psi}{\partial x}$$

It should also be noted that based on this definition the functions ϕ and ψ are orthogonal to each other and it is possible to use one or the other to represent the flow.

A new complex function can be defined, whose real and imaginary parts are based on the velocity potential and streamfunction as:

$$F = \emptyset + i\psi \tag{5.31}$$

The derivative of F in terms of z is defined as:

$$W(z) = \frac{dF}{dz} \tag{5.32}$$

Also:

$$\frac{dF}{dz} = \frac{dF}{dz}\frac{\partial z}{\partial x} = \frac{\partial F}{\partial x}$$

Consequently, inserting the definition of F we have

$$W(z) = \frac{\partial \emptyset}{\partial x} + i\frac{\partial \psi}{\partial x} = u - iv \tag{5.33}$$

This result shows that $W(z)$ represents the "complex velocity" of the flow and is determined by the real velocity components u, v.

It is often advantageous to use cylindrical coordinates, expressed as r, θ in two dimensions. The transformation from x, y to r, θ is:

$$u = u_r cos\theta - u_\theta sin\theta$$

$$v = u_r sin\theta + u_\theta cos\theta$$

The expression for the complex velocity is then (by direct substitution and using the identity of $\cos\theta - isin\;\theta = e^{-i\theta}$:

$$W = (u_r - iu_\theta)\, e^{-i\theta} \tag{5.34}$$

The use of the complex variable representation in terms of F can then be converted back into the physical space velocity components, u, v, through its derivative relative to the variable z. The magnitude of the local velocity vector is found from W by taking the square root of $WW*$, where $W*$ is the complex conjugate of W. Once the velocity is determined then the pressure field is directly determined from the Bernoulli Equation.

Some examples will help illustrate the use of complex variables to represent rather simple flows.

Uniform Flow

A uniform flow of magnitude U in the x direction becomes:

$$\emptyset = Ux = Ur\cos\theta \;and\; \psi = Uy = Ur\; sin\theta$$

$$F = Ux + iUy = U(x + iy) = Uz$$

$$W = u - iv = \frac{dF}{dz} = U$$

Note that for a flow, V, in the y direction F is equal to $-iVz$.

By tilting the uniform flow (direction of U) by angle α relative to the x axis the general expression for F in terms of z becomes:

$$F = Uze^{-i\alpha}$$

$$u = Ucos\ \alpha; v = Usin\ \alpha$$

Source/Sink Flow

A source and sink flow is radial flow from a point and has a strength proportional to its volumetric flow rate such that for any circle centered about the origin of the source or sink, the line integration about the circle yields the volume flow rate per depth into the plane of the circle (recall that we are only dealing with two dimensional flows). A source has a positive strength with flow outward from the center and a sink has a negative strength (flow is inward towards the center). If the volume flow rate per depth is Q' then the strength is designated as $\mu_s = Q'/2\pi$. The equations governing the velocity potential lines and streamlines are:

$$\emptyset = \mu_s \ln r\ and\ \psi = \mu_s \theta$$

The resulting expression for the complex potential is:

$$F = \mu_s \ln r + i\ \mu_s\ \theta = \ \mu_s (\ln r + i\ \theta) = \mu_s \ln \left(re^{i\theta}\right) = \mu_s \ln z$$

To locate a source or sink at a point other than the origin, say at location zo, we have:

$$F = \mu_s \ln(z - z_o)$$

Vortex Flow

A vortex flow is one with only a circumferential velocity component about the origin. The velocity decays proportional to $(1/r)$, where r is the radial coordinate. Notice that the resultant streamlines and potential lines for this flow are orthogonal to the streamlines and potential lines for a source flow. The integral of the velocity around a closed path that includes the origin is defined as the circulation, Γ. For convenience we chose a circular path of radius r:

$$\Gamma = \oint u \bullet dl = \oint v_\theta r d\theta = 2\pi C$$

Where $C = v_\theta r$ which is a constant since v_θ varies as $1/r$. We define $C = \mu_v$ as a measure of the strength of the vortex as

$\mu_v = \Gamma/2\pi$

So, noting that velocity potential lines are radial and streamlines are circular (counterclockwise) as:

$$\emptyset = \mu_v \theta$$

$$\psi = -\mu_v \ln r$$

$$F = \mu_v \theta - i\,\mu_v \ln r = \mu_v\left(\theta - i\,ln\,r\right)$$

$$F = \; -i\mu_v \ln re^{i\theta}$$

Lastly, the complex velocity can be written as:

$$W\left(z\right) = -i\frac{\mu_v}{z} = -i\frac{\mu_v}{r}e^{-i\theta}$$

The negative sign for this expression for W results in a rotation in the counterclockwise direction, which is typically selected as "positive" rotation. Notice the resemblance to the formulation for the source flow and vortex flow representation keeping in mind the orthogonal condition of $\emptyset$ and ψ . It would be a good exercise to generate plots of the above equations for velocity potential and streamfunction.

There may be some concern that a vortex, which contains vorticity can be considered as irrotational flow. This can be explained as follows. If one were to form a contour anywhere in the flow that does NOT contain the origin (the location of the vortex center) then it can be shown that the circulation is zero. One could conclude that all of the vorticity is located at the center, representing a singularity. The flow driven by this singularity is indeed irrotational. Another way to establish this is to form two contour circles both with centers at the vortex origin but of different radii. It is easily shown then that the difference of the circulation between these two is zero, concluding that for arbitrarily selected contour circles there is no rotational flow between them – so the flow must everywhere be irrotational except at the center.

Flow in a Sector

A sector is defined here to be a region in space near the intersection of two lines, as in a corner with an arbitrary angle. There are general formulations for flows in the region of a sector that can be written in as:

$$F = Az^n = Ar^n\left(\cos n\theta + i\sin n\theta\right) \tag{5.35}$$

Where "n" represents a parameter to be specified, usually as a constant for a given flow, and "A " is a constant for a specific flow, and shown below to be proportional to the velocity far from the sector, representative of the flow into the sector. An example of this class of flows is uniform flow where $A = U$ and $n = 1$. Note that uniform flow can be thought of as flow over a flat surface that is parallel to the uniform flow direction (there are no viscous forces so the no slip boundary condition does not hold). The flat surface can be thought of as a sector with an intersection of two lines with the lines being parallel. Or, the angle between the two lines is π. The general expression above corresponds to the following expression for velocity potential and streamfunction:

$$\emptyset = Ar^n \cos(n\theta)$$

$$\psi = Ar^n \sin(n\theta)$$

Using these two expressions we can easily locate lines of constant streamfunction can be found, and in particular when $\psi = 0$: when $\theta = 0$ and π/n. The flow between these two radial lines represents flow in a "sector", as seen in the Fig. 5.8, below.

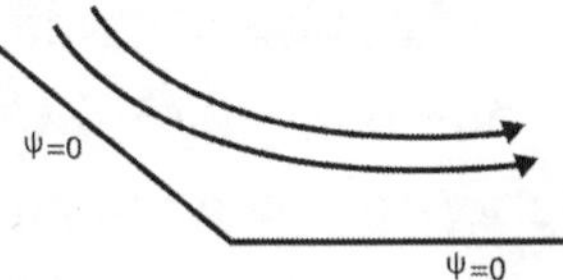

Fig. 5.8 Two-dimensional flow in a sector of arbitrary angle between two straight lines; the streamfunction is chosen to be zero along these two lines.

Using the definition of the complex velocity potential, W:

$$W = \frac{dF}{dz} = nAz^{n-1} = nAr^{n-1}e^{i(n-1)\theta}$$

Expanding the exponential reveals that the velocity components in r, θ coordinates are:

$$u_r = nAr^{n-1} \cos(n\theta)$$

$$u_\theta = nAr^{n-1} \sin(n\theta)$$

By inserting different values for the parameter "n" different flows can be simulated. Some examples are given below.

$n = 1:$
$u = A$ and $v = 0$ (uniform flow)

$n = 2:$

$u = 2Ar\cos(\theta)$
$v = 2Ar\sin(\theta)$
or in Cartesian coordinates: $u = 2Ax;\ v = 2Ay$
$\psi = -Ar^2 \sin(2\theta) = -Axy$

or for a constant streamfunction (along a streamline) we have:

$$xy = C$$

(where C is a constant determined by the value of ψ at a given location). This latter flow, for a range of values of C yields flow into a right angled corner where the location $(0, 0)$ is the position of the corner. Note that $\psi = 0$ along the axes. The reader is encouraged to plot this function for different values of streamfunction to visualize the flow.

The corresponding flow into a corner of arbitrary angle, α, can be expressed as:

$$F = Az^{\pi/\alpha}$$

Note that Az^n is equivalent to $Ar^n\left[cos(n\theta) + i\ \ sin(n\theta)\right]$ which can then be restated in terms of $\pi/\alpha = n$, so that for $\alpha = \pi/3$ we have $n = 3$. This then relates n to the turning angle of the corner flow. For instance, if $n = 3/2$ we have an angle of $2\pi/3$ which is greater than 90°. We can also have flow over a flat plate parallel with the flow if $n = 1$.

A wedge flow for $\alpha > \pi$

Wedge flow resembles the flow that divides at the front side (or leading edge) of an object and is a reasonable model for the flow over the leading portion of certain objects. This is illustrated in Fig. 5.9 where flow impinges on an object with a stagnation point at the wedge vertex and then accelerates along the top and bottom surfaces as the streamlines converge. With $\alpha > \pi$ then $n < 1$.

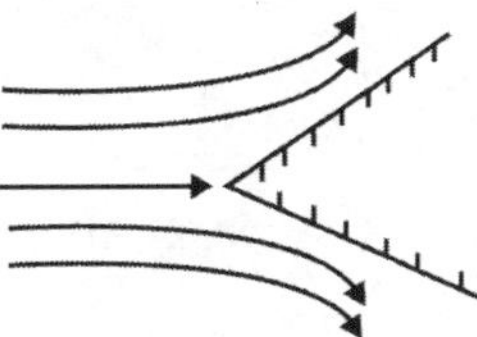

Fig. 5.9. Illustration of a wedge flow with a stagnation point at the vertex; along the top and bottom surfaces the flow accelerates resulting in lower pressures on either side of the wedge.

An extreme for wedge flow is flow around a sharp edge, shown in Fig. 5.10 below. Here the flow is moving parallel with a surface, reaches a sharp edge of the surface, and then flows around the edge and then back parallel with the surface. The streamfunction is typically given the value of zero on the surface, and the surface is infinitely thin and flat.

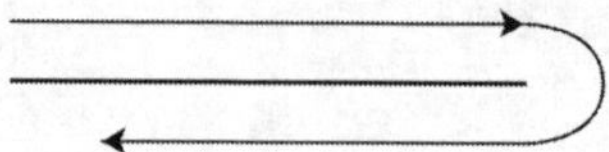

Fig. 5.10. Illustration of flow over a flat surface with a trailing edge.

This flow is represented mathematically in terms of a $F(z)$ and constant, $A, n = 1/2$ as:

$$F\left(z\right) = A\; z^{1/2}$$

$$F\left(z\right) = \;A\, r^{1/2} e^{i\frac{\theta}{2}}$$

$$\psi = Ar^{1/2}\,\sin\frac{\theta}{2}$$

This expression for the streamfunction can be shown to be equal to zero along the surface for $\theta = 0, 2\pi$. The velocity is obtained from the streamfunction to be:

$$u_r = \frac{1}{r}\frac{\partial\psi}{\partial\theta} = \frac{A}{2r^{1/2}}\cos\frac{\theta}{2} \qquad u_\theta = \; - \frac{A}{2r^{1/2}}\sin\frac{\theta}{2}$$

it can be verified that for $0 < \theta < \pi : u_r > 0$ and $u_\theta < 0$; and $\pi < \theta < 2\pi : u_r < 0$ and $u_\theta > 0$. This assures the flow reversal around the edge.

Using the flows above written in terms of the constant A we have assumed A to be a real number, but this is not required. We can express this as a complex number by including $e^{-i\beta}$ as a new term: $Ae^{-i\beta}$, where in this expression A is a real number and β is a constant not the same as α used above to define the turning angle of the flow. The result is

$$F = Ae^{-i\beta}\,z^n = Ar^n\,e^{i(n\theta-\beta)} \tag{5.36}$$

Noting that the real part of F is the velocity potential and the imaginary part is the streamfunction we have:

$$\emptyset = Ar^n\,\cos(n\theta - \beta)$$

$$\psi = Ar^n\,\sin(n\theta - \beta)$$

This streamfunction expression shows us that when using the complex representation for the constant multiplying z^n there is introduced a rotation of the streamlines through angle β using the same coordinate system when compared to the streamfunction when the coefficient is real (not complex).

Doublet

Next, we introduce the concept of a doublet using complex notation (the superposition of a source and sink brought together in the limit as the separation distance goes to zero – but never actually reaches zero separation). This can be expressed as:

$$F = \frac{\mu_D}{r} e^{-i\theta}$$

$$\emptyset = \frac{\mu_D}{r} \cos\theta$$

$$\psi = \frac{\mu_D}{r} \sin\theta$$

where μ_D is a constant representing the strength of the doublet. Lines of constant streamfunction $\psi = B$ become

$$\psi = B = \frac{-\mu_D \; r \; sin\theta}{r^2} = \frac{-\mu_D y}{(x^2 + y^2)}$$

$$x^2 + (y + \frac{\mu_D}{2B})^2 = \left(\frac{\mu_D}{2B}\right)^2$$

This last equation illustrates that we have an equation of a circle with the center located along the y axis with the position changing depending on the value of B, which implies different streamfunction values and therefor different streamlines for each B. The radius of each circle is determined by $(\mu_D/2B)$ which depends on the value of the streamfunction. Also, each circle is tangent to the x axis at $x = 0$, or the line with $\theta = 0$ at $r = 0$. See the sketch below in Fig. 5.11 of streamlines.

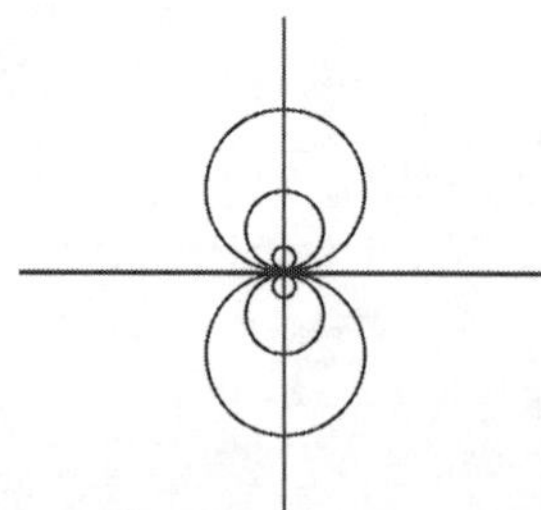

Fig. 5.11. Illustration of streamlines associated with a doublet.

Rankine Half-Body

More complicated flows using complex variable representation based on the superposition of simpler flows can be formed with any proper representation of the simpler flows. An example is the Rankine half-body. In some ways this has characteristics similar to the wedge flow identified above in that we are interested in the flow approaching an object and the initial region near the leading edge. However, rather than flat plates forming a corning with a sharp point we have the simulation of flow over a rounded leading-edge surface. In this case we superimpose a uniform flow with a source located at the origin of the coordinate system, both expressed using complex variables. Here we have:

$$F = Uz + \mu_s \ln z$$

To obtain the velocity components in the (r, θ) coordinates we use the definitions of the velocity potential (or we could use the streamfunction):

$$u_r = \frac{\partial \emptyset}{\partial r} = U cos\, \theta + \frac{\mu_s}{r}$$

$$u_\theta = \frac{1}{r}\frac{\partial \emptyset}{\partial \theta} = -U sin\, \theta$$

At the very front, or leading edge of the body, the velocity should become zero, representative of a stagnation point. So, setting each velocity component equal to zero we can determine where the stagnation point should occur.

$$u_r = 0: \quad U cos\, \theta = -\frac{\mu_s}{r}$$

$$u_\theta = 0: \quad U \sin \theta = 0$$

There are two solutions to the above set of equations: $\theta = 0$ and $\theta = \pi$. However, for $\theta = 0$ the value of r becomes negative from the first equation, which is an invalid condition so the only realistic solution for the location of the stagnation point is $(\mu_s/U, \pi)$. By increasing the source strength, $\mu_s v$, the stagnation point moves further upstream in the flow. The streamfunction that passes through this point is found by noting that $u_\theta = -\frac{\partial \psi}{\partial r}$ such that integrating and setting $\theta = \pi$ then $\psi_{stag} = \mu_s \pi$. Since the streamfunction passing through the stagnation point is on the body then a general streamfunction expression for the body becomes

$$\psi_{body} = U r_{body} \sin \theta_{body} + \mu_s \theta_{body}$$

In this expression the subscript "body" represents r, θ coordinates that are on the body which will be a constant value of the streamfunction. A plot of a constant value of $\psi_{body} = \mu_s \pi$ (which is specified by the stagnation point given above) can be used to determine the body surface. In general, we have:

$$r_{body} = \frac{\mu_s (\pi - \theta_{body})}{U \sin \theta_{body}}$$

One can imagine that for increasing r in the first or fourth quadrant the influence of the source will diminish (since we are moving away from the source). Consequently, the flow is expected to eventually become streamlines that are again parallel with the oncoming flow. For the streamline of the body, in the limit of large r we can write an expression for the total width of the body, w_{body} :

$$w_{body} = \frac{2\pi \mu_s}{U}$$

By increasing the source strength for a given freestream velocity U the width of the body increases and it is possible to model flow over the object of any desired width.

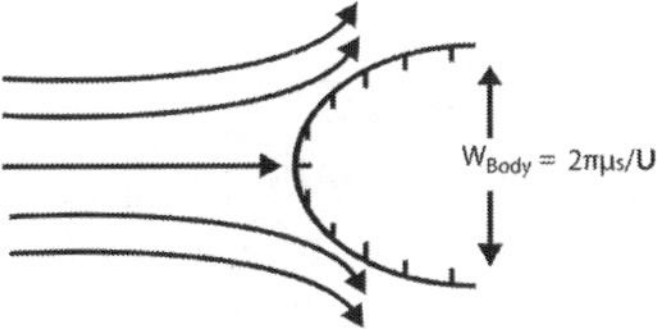

Fig. 5.12. Flow over a Rankine half-body.

Enclosed Bodies

It is possible to continue on with the idea of flow over an object and create a completely enclosed object (this requires an enclosing streamline such as a circle or ellipse) to simulate flow over the enclosed object. The streamlines inside the body are of no interest here, just those flowing around the outside of the body and how the velocity and pressure vary. As an example, we simulate the flow over a circular cylinder. This will involve the superposition of a uniform flow and a doublet at the origin. This can be expressed in our complex representation as:

$$F(z) = Uz + \frac{\mu_D}{z}$$

The strength of the doublet for a given flow U will determine the radius of the cylinder. Consider a desired radius a for a given uniform flow U. We can write the complex potential for the circle as:

$$F(z) = Uae^{i\theta} + \frac{\mu_D}{ae^{i\theta}}$$

Noting that the streamfunction is the imaginary part of this expression which yields:

$$\psi = \left(Ua - \frac{\mu_D}{a}\right)\sin\theta$$

We can force the value of the streamfunction to be $\psi = 0$ at $r = a$ by setting the strength of the doublet to $\mu_D = Ua^2$. Consequently, we end up with flow of velocity U over a cylinder of radius a. We insert the strength of the doublet into the expression for $F(z)$:

$$F = U(z + \frac{a^2}{z})$$

Again, we are only interested in the flow outside of the cylinder, as shown in Fig. 5.13.

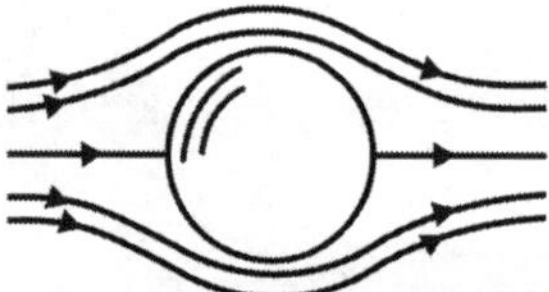

Fig. 5.13. Flow over a sphere; note symmetry of streamlines about the x axis and y axis.

As we will see later, it is possible, through a proper transformation of the flow over a cylinder, to obtain the flow over objects, such as an airfoil. However, to make this more physically correct we will want to introduce circulation to the flow to simulate the circulation that occurs for an actual airfoil that provides the lift force experienced by the airfoil. Recall from a vortex flow that circulation is based on circumferential flow, with strength proportional to the u_θ velocity component. Superimposing a uniform flow over a cylinder with radius a, with a vortex of strength μ_v rotating in the clockwise direction, results in the following:

$$F = U\left(z + \frac{a^2}{z}\right) + i\mu_v \ln z + c$$

where we add the constant c, this will allow us to assign the value of the streamfunction to the cylinder surface, at $r = a$. By setting $z = ae^{i\theta}$ to be on the cylinder surface the value of the complex potential can be evaluated and thereby obtaining the streamfunction value in terms of the unknown constant c. By setting $c = \ - i\mu_v \ln a$ it is straightforward to show that the streamfunction is in fact equal to zero for all θ at $r = a$. Then using this value for c and combining c with the second term (combining the ln terms) results in the following expression for the flow over a cylinder with circulation:

$$F = U\left(z + \frac{a^2}{z}\right) + i\mu_v \ln\frac{z}{a}$$

Note that the above expression is for circulation with clockwise rotation, for counterclockwise rotation the ln term is negative.

The velocity field associated with the above expression for $F(z)$ is:

$$u_r = U\left(1 - \frac{a^2}{r^2}\right)\cos\theta$$

$$u_\theta = U\left(1 - \frac{a^2}{r^2}\right)\sin\theta - \frac{\mu_v}{r}$$

Notice that it is the vorticity that provides the circumferential flow component. Also, along the surface of the cylinder the velocity is not necessarily zero. In fact, since this is a streamline then the value of u_r must be zero so no flow crosses the cylinder, but u_θ is only zero at two points. These points are stagnation points because both velocity components are zero resulting in zero velocity. The locations of these two points are found by setting $u_\theta = 0$ and solving for θ:

$$\sin\theta_{stagnation} = \frac{\mu_v}{2Ua} = \frac{\Gamma}{4\pi Ua}$$

If the circulation is zero ($\mu_v = 0$) then the stagnation points are at $\theta = 0, 2\pi$, implying symmetric flow both relative to the x and y axis. See Fig. 5.14 for representation of stagnation points for different circulation conditions. For clockwise rotation greater than zero stagnation points will move to the third and fourth quadrants. For counterclockwise rotation, the stagnation points will be in the first and second quadrants. When the strength of the clockwise circulation is large enough to result in $\theta = 3\pi/2$ and the two points coincide on the "bottom" of the cylinder. For larger circulation strengths than this the stagnation point actually moves off of the cylinder.

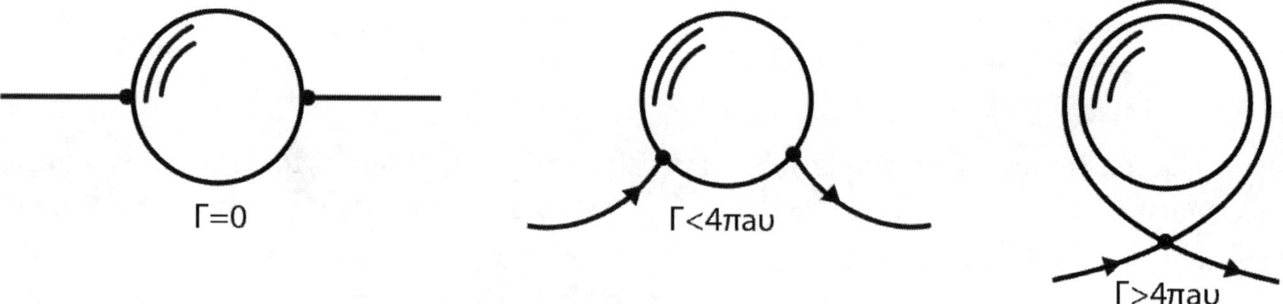

Fig. 5.14. Stagnation points on a cylinder with different circulation, [latex]\Gamma[/latex].

At this point it is instructive to evaluate the consequences of the added circulation to flow over a cylinder. As we have seen from above, the circulation removes symmetry across the x axis within the flow. That is the streamline distribution for the third and fourth quadrants is different from the first and second quadrants of the flow. However, this circulation does not change the symmetry that exists across the y axis (the flow in the first and fourth quadrants is symmetric with the second and third quadrants). Recall from the Bernoulli equation (viscous forces are ignored and we have a steady flow) that the pressure field will retain a similar symmetry with the velocity field. The symmetry across the y axis indicates the pressure on the "front" half of the cylinder will be identical to that on the "back" half of the cylinder and the net force in the x direction will be zero. However, across the x axis the loss in velocity symmetry implies a difference in pressure distribution on "top" of the cylinder versus the "bottom". Also, for circulation with a clockwise rotation the velocity on top will be greater than that on the bottom resulting in lower pressure on the top. The net effect is an upward force on the cylinder by the pressure field. If the rotation is reversed there would be a net downward force on the cylinder.

The determination of the force on a body with external flow around it in potential flow can be determined from the Kutta-Joukowski law which relates the net lift on the body to the circulation, Γ, generated by the flow and the magnitude of the freestream velocity, U, as:

$$Lift/span = -\rho U \Gamma$$

This relationship is shown in an earlier part of this chapter. It is also shown below in the context of conformal mapping. In this formulation, if the circulation is negative (clockwise) then the force direction will be positive (upward lift) when there is positive flow over the body (left to right). This can be derived for any shaped two-dimensional body with an associated circulation calculated from the line integral about an enclosed area that includes the body. This law can be shown through the use of Newton's law relating total force (lift plus pressure) to rate change of momentum and the integration of the pressure field obtained from the Bernoulli equation for steady two-dimensional, irrotational inviscid flow. We show this also in the next section. The interested student can review the derivation of this from many references or online using the basic tools of complex representation of the velocity field, for instance: https://en.wikipedia.org/wiki/Kutta–Joukowski_theorem.

For the circular cylinder case the circulation is specified within the strength of the vortex and is used in conjunction with the freestream velocity to specify the flow conditions associated with a given freestream velocity and cylinder radius.

Conformal Transformations

As mentioned previously it is possible to transform the results for flow over a cylinder with circulation to that of flow over an airfoil, allowing the determination of the lift force through the solution to the magnitude of the lift force for flow over the cylinder. The idea of a transformation in this sense is that by altering the coordinate system it is possible to change from a circular geometry to a different geometry. Once this transformation is specified it is possible to calculate the flow at certain points within the circular geometry and assign the flow results to points in the new, different geometry.

Consider a mapping function $\zeta = \zeta(z)$ which establishes a relationship between coordinates ζ and z (where again $z = x + iy$). Consequently, a known solution expressed in ζ can then be transformed into the z coordinate. The condition for conformal mapping in this sense is that the velocity potential (which satisfied the Laplace equation in the original coordinate must also be satisfied in the new coordinate. The other condition is that new coordinate must satisfy the Cauchy-Riemann equations. These state the following where the new coordinates are, say, (ξ, η) and the original coordinates are (x, y):

$$\frac{\partial \xi}{\partial x} = \frac{\partial \eta}{\partial y} \; and \; \frac{\partial \xi}{\partial y} = -\frac{\partial \eta}{\partial x}$$

Under these conditions it can be shown that the complex potential, $F(z)$, in a given coordinate which determines the velocity potential and streamfunction, can be shown to satisfy the Laplace equations for the velocity potential and streamfunction in the ζ plane. The local solutions in the z plane are found from the solutions in the ζ plane by using the relationship given by $\zeta = \zeta(z)$. In other words, a solution at a point in ζ has the same value at the corresponding point in z. In additional to this, if we take our previous definitions for $W(z)$ we can write:

$$W(z) = \frac{dF(z)}{dz} = \frac{d\zeta}{dz}\frac{dF(\zeta)}{d\zeta} = \frac{d\zeta}{dz}W(\zeta)$$

This says that the mapping function derivative with respect to z defines the relationship between the complex velocity solutions in each coordinate frame. Lastly, it can also be shown that the strengths of the various flow elements used within one frame are not altered in the transformed coordinate frame. These relationships now allow one to solve a given problem, such as flow over a cylinder with circulation, and interpret the results in another coordinate frame which may change the geometry of the object.

As an example, consider using the Joukowski transformation:

$$z = \zeta + \frac{C^2}{\zeta}$$

with C being a constant that determines the shape of the boundary in transformation. For instance a circle of radius a given by $\zeta = ae^{i\theta^*}$ while setting $C = a$ results in:

$$z = a\left(e^{i\theta^*} + e^{-i\theta^*}\right) = 2a\cos\theta^*$$

and the resultant z plane shape will be a flat plate extending from $-2a$ to $+2a$ along the x axis in the z plane. Note that here, $\theta*$ is used to represent the angle coordinate in the ζ plane (not z plane). This illustrates a transformation between a circle, in ζ, to a flat plate in z. Here we are interested in points outside of the radius of the circle and how they transform to points in the z plane.

Continuing this, by noting that the stagnation points in the ζ plane are determined by the circulation, as shown previously for flow over a cylinder, then when the circulation is zero the stagnation points are at $\theta* = 0, \pi$. The stagnation points in the z plane then are found through the transformations as at $x = \pm 2a$, or at the ends of the flat plate.

The more interesting case is when we have a flat plate with the freestream flow at some angle, a, to the x axis. This is usually defined to be the angle of attack. When this occurs a stagnation point will form on the "bottom" side of the plate when the freestream velocity has a positive angle relative to the x axis (which forms the freestream velocity direction). This is shown in fig. 5.15, below. At this stagnation point the flow separates, with part of the flow moving forward in the positive x direction and some of the flow moves in the negative x direction. The latter flow then must turn around the corner or edge of the flat plate as shown in the figure. Also, the trailing edge will have a relative movement of the stagnation point away from the trailing edge of the plate, resulting in flow having to go around the trailing edge and up along the surface. Since this flow is inviscid the potential flow velocity around an infinitely thin plate must go to infinity since the radius of curvature is zero. This does not happen in a real flow due to viscous effects slows the fluid and resulting in a finite radius of curvature and flow separation at the edge. In a real airfoil the leading edge has a finite thickness and can potentially eliminate flow separation unless the angle of attack is too large. But the trailing edge is typically very thin. The resulting flow condition preventing separation at the trailing edge is the Kutta condition imposed at the trailing edge. See the earlier discussion of the Kutta condition. The idea is that the flow adjusts itself to prevent this separation by imposing local circulation near the trailing edge to offset the separation flow. If circulation is added to the flow to move the stagnation point to the trailing-edge then the flow at the trailing edge will have equal velocity on the top and bottom surfaces of the plate (and by Bernoulli equation equal pressures) such that the flow will leave the trailing edge smoothly and not wrap around the trailing edge.

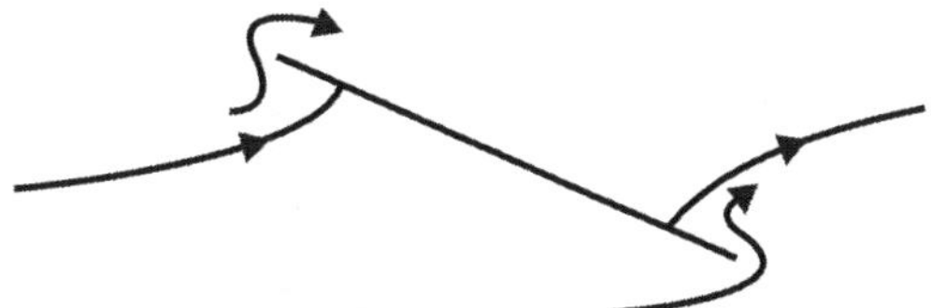

Fig. 5.15. Stagnation points for flow over a flat plate at a nonzero angle from the freestream flow.

Considering flow around a cylinder in the ζ plane at an angle of attack of α relative to the x axis, or $\theta = 0$. The uniform freestream must be rotated by angle α, and this results in the following representation of the complex potential:

$$F(\zeta) = U\left(\zeta e^{-i\alpha} + \frac{a^2}{\zeta e^{-i\alpha}}\right) + i\mu_v\, ln\frac{\zeta}{a}$$

However, we wish to impose the Kutta condition as well. To do this we must add circulation equivalent to moving the stagnation point to the trailing edge, or $z = 2a$. For this value of z the corresponding value of ζ is a, which is on the cylinder at $(a, 0)$. To achieve this additional rotation of the flow by angle α in the ζ plane is required. We use the equation for the stagnation point position, $\theta_s tag$, to see that the required circulation is $\Gamma = 4\pi U a sin\ \alpha$.
So inserting this value of circulation into the vortex strength above we obtain:

$$F(\zeta) = U\left(\zeta e^{-i\alpha} + \frac{a^2}{\zeta e^{-i\alpha}}\right) + i2Ua\ sin\alpha\ ln\frac{\zeta}{a}$$

This can be written in terms of the z plane with $\zeta = \frac{z}{2} + \sqrt{\left(\frac{z}{2}\right)^2 - a^2}$

as:

$$F(z) = U\left(\left(\frac{z}{2}+\sqrt{\left(\frac{z}{2}\right)^2-a^2}\right)e^{i\alpha}+\frac{a^2e^{i\alpha}}{\left(\frac{z}{2}+\sqrt{\left(\frac{z}{2}\right)^2-a^2}\right)}+i2a\ sin\alpha\ ln\left(\frac{1}{a}\left(\frac{z}{2}+\sqrt{\left(\frac{z}{2}\right)^2-a^2}\right)\right)\right)$$

Noting that the imaginary part of $F(z)$ represents the streamfunction. Fig. 5.16 below represents the flow in both the z and ζ planes.

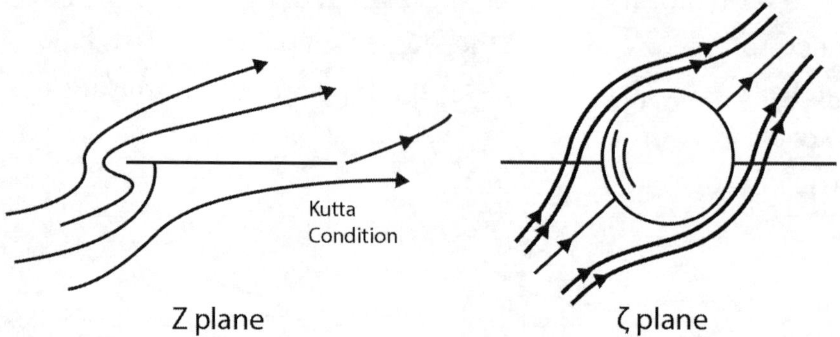

Fig. 5.16. Transformation between flow over a flat plate and flow over a cylinder.

We can easily determine the lift force with the known circulation to be:

$$L = \rho U\Gamma = 4\pi\rho U^2 a \sin\alpha$$

Expressing this nondimensionally, as a lift coefficient by dividing by $(1/2\rho U^2 c)$ where c is the length of the plate (the chord length) which is equal to $4a$ we obtain:

$$C_L = 2\pi\sin\alpha$$

This is the theoretical lift coefficient for a thin (flat) airfoil at an angle of attack of α, as was shown previously in this chapter. For small angles of attack, we have $C_L = 2\pi\alpha$.

Blasius Theorem and Lift Force for an Arbitrary Body

The lift force for a rotating cylinder in a uniform inviscid flow is given by $L = -\rho U\Gamma$ (where the circulation is positive if counterclockwise and flow is from left to right). What is rather amazing is that the object shape is not critical in the use of this same equation as long as the value of the generated circulation is determined properly.

The Blasius Theorem (also referred to as the Blasius Integral Laws), is a means to obtain the total force on the object within a flow. Ideally the surface velocity distribution would be found, then using the Bernoulli equation the pressure along the surface is found, which could be integrated to find the net force on the object. However, this procedure can be circumvented through the use of the Blasius Theorem.

At any location on the object surface the expression for the local force, dF, in terms of the velocity potential, can be found. Specifically, the force is decomposed into drag and lift components, noting that drag is in the negative x direction, as:

$$dF = dF_D - idF_L = -Pdy - iPdx = -iP(dx - idy)$$

Now to extend this around the entire object it is needed to integrate around the closed path forming the object (by sign convention we go counterclockwise) as:

$$F_D - iF_L = -i \oint Pdz^*$$

$$z^* = x - iy$$

Where $z*$ is the complex conjugate of z. The pressure is determined from the Bernoulli equation with the surface velocity written as

$$U_s{}^2 = [(u + iv)(u - iv)]_s$$

$$u + iv = (u_r{}^2 + u_\theta{}^2)^{1/2} e^{i\theta}$$

$$dz = |dz| e^{i\theta}$$

The last two expressions show that the velocity and surface are parallel with each other, both at the same angle, as must be the case for a solid surface in inviscid flow. So, we can show that $(u + iv)dz*$ is a real number (using the fact that the velocity slope equals the surface slope, or $v/u = y/x$, so the imaginary part goes to zero) and that $(u + iv)dz* = (u - iv)dz$. Finally, the velocity is determined by:

$$W = (u - iv)$$

So, by inserting for the pressure in terms of the velocity from the Bernoulli equation, and using the above express for U_s along with the relationship for W in terms of the velocity components it can be show after a bit of manipulation that:

$$F_x - iF_y = \frac{i\rho}{2} \oint W^2 dz$$

The above expression is valid for steady, potential flow and often called the "Blasius Theorem". Notice that the drag component is the negative of the real part of the right hand side and the lift component is the negative of the imaginary part.

In potential flow the integration around any closed contour (say a contour around the surface of a body versus a contour around the body far from the body itself) can be shown to be the same. For this to be true there can be no singularities that occur between the contours. So, in this case no sources or sinks or vortices, etc. can exist between the surface contour and a contour far from the surface. Flow around an arbitrarily shaped object may be generated by putting a distribution of singularities, such as vortices, around the shape contour and adjust their strength distribution to form a closed contour of some desired shape. This is discussed in detail in the next chapter. In the flows considered here there are no singularities outside of the body itself. So, we can take a contour far from the body out into the free stream. All of the distributed singularities will appear, from far away, as if they are all located at the coordinate system origin (assumed to be located near or in the object-although this may not be necessary). The superposition representation of the complex potential can be written by eliminating the small distances from singularity location and the origin. One can then insert the complex velocity associated with the superposition into the above equation for force. The unique aspect here is that as the contour moves further and further away from the body the complex velocity can be simplified by dropping terms that are small, such as for a vortex we have terms like $i\Gamma/2\pi z$ and for a doublet we have μ_D/z^2. Note that here a positive circulation is designated as clockwise, as is typically done in aerodynamic applications. If we only retain terms ~$1/z$ and perform a series expansion then by complex variable theory the contour integral is equal to $2\pi i \sum residuals$.

The residuals are determined by taking the square of the complex velocity as shown in the integral above, and dropping terms higher than say $1/z$. Then the residual is the coefficient of the $1/z$ term which is multiplied by $2\pi i$ to obtain the value of the integral. The real part is F_D and the negative of the imaginary part is F_L.

This can be illustrated using the complex potential for a uniform flow, vortex and doublet (flow over a rotating cylinder): $F = Uz + \mu_v \ln z + \dfrac{\mu_D}{z}$. Taking the z derivative to form the complex velocity, W, and then squaring this and only retaining up to the $1/z$ terms, the coefficient of the $1/z$ term will be $2i\mu_v$ resulting in the following for the force:

$$F_D - iF_L = \frac{i\rho}{2}\left(2\pi i \frac{(i\Gamma U)}{\pi}\right) = 0 - i\left(\rho U \Gamma\right)$$

This states the there is a net lift force dependent of the circulation and zero drag force and is the identical to the result presented previously. The circulation here is that due to the sum total effect of the distribution of vortices around the object. This result is often called the Kutta-Zhukhovsky theorem.

VI. The Panel Method: An Introduction

The panel method is an analysis method that can be used to arrive at an approximate solution for the forces acting on an object in a flow. The method, as we present it here, is based on inviscid flow analysis, so it is limited to the resultant pressure forces over the surface. The panel method is basically a numerical approximation that relies on using discrete elements on the surface of an object and then prescribing a flow element (such as a vortex or doublet or source or sink) on each element that will satisfy certain boundary conditions (like no flow crosses the surface of the object). The interaction of the elements are accounted for and must also satisfy the condition that far from the object the flow should be equal to the free stream velocity approaching the object. There are a number of books and papers written that describe the method in very general terms and even the inclusion of viscous forces to some degree. But here we are just introducing the method to get a feel for its usefulness in external flows, so we will use a simply geometry with a simply distribution of flow elements. More complicated models exist but they all are based on the simplified form presented here.

We will assume that we have potential flow such that the governing equation for the flow field is the Laplace of the velocity potential, $\nabla^2\phi = 0$. The boundary condition at an impermeable surface, where the velocity normal to the surface is zero, is $\nabla\phi \cdot \boldsymbol{n} = 0$. Also, we can put our frame of reference on the object so fluid flow approaches the object. Keep in mind that since it is inviscid there may be a nonzero velocity component tangent to the surface. Also, the goal is to determine the velocity on the surface, and once this is found the Bernoulli equation can be used to find the local pressure distribution. The pressure can then be integrated over the surface to find the force by the fluid flow.

Without deriving this it can be shown that the following defines the velocity potential at any point P in the flow field (using Green's Identify):

$$\phi\,(P) = \frac{1}{4\pi}\int(\frac{\nabla\phi}{r} - \phi\nabla\frac{1}{r})\cdot ndS \tag{6.1}$$

where the integral is over the surface area of the flow (assuming two dimensional flow), S. This equation indicates that to solve for the velocity potential we must evaluate the integral on the flow boundaries (both the solid surface and infinitely far away).

This can be shown to be a result of the application of the divergence theorem that says:

$$\int n_j u_j dS = \int \frac{\partial u_i}{\partial x_i} d\forall \tag{6.2}$$

where u_i is the velocity vector. This states that the rate of expansion of a volume of fluid, given by the divergence of the velocity vector, within a given volume equals the flux through the volume boundaries. If the fluid is incompressible then the right side is zero and the net flux through the entire boundary is zero (this is the general argument for the conservation of mass of steady flow).

All this is nice and can be a powerful tool to find ϕ and therefore the velocity in the flow. But we really don't need to find the potential of the entire flow, what our goal is, is to set up a flow that we know satisfies the velocity boundary condition at the surface of the object (with no velocity component across the surface) and far from the surface where the velocity is known to be the freestream velocity. Once this is established the force can be found. That is to say we only need to evaluate the surface velocity and then the pressure on the surface.

The general approach is to select a "grid" which is a series of "panels" that form the surface. Here we take the panels as straight flat surfaces arranged over the real surface. In the limit of very small panels the constructed surface will simulate the actual surface. On each panel we place a distribution of flow elements (like sources, sinks, vortices, etc) that when combined together will result in a flow field that will satisfy the surface boundary condition. There are lots of ways to identify which elements to use and how they may be distributed on the panels. Here we will use vortex elements, with one placed on each panel. The net flow is the result of superposition of the flow set up by each vortex on each panel element. So at each point in the field we add together thc flow caused by all of the panel elements using the superposition rule. The panels that are far away from a given point will have less and less influence on the flow because the strength of the flow caused by a flow element decreases with distance from the element origin. For instance for a "source" the velocity decreases with increasing radial position because the flow is spreading out away from the source. But the influence never really goes to zero.

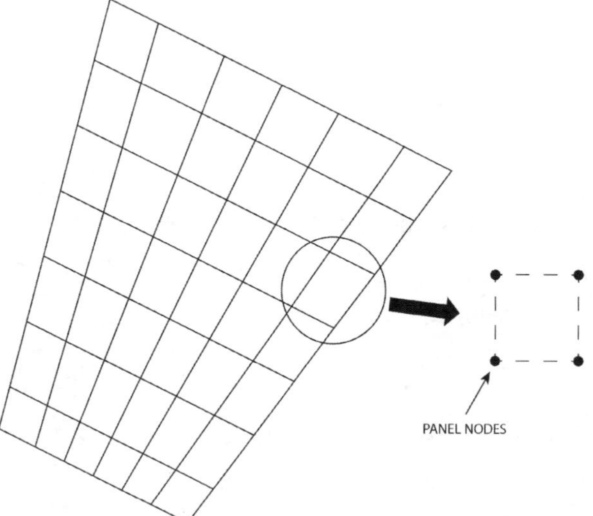

Fig. 6.1 Illustration of a panel geometry; any three dimensional shape can be constructed; shown here is a surface of what could be a three dimensional object such as an entire airplane.

In placing a series of panels over the surface we first need to specify the size of each panel. We place a vortex of some strength some where on the panel (whose location is shown below) and we must identify points on the surface where we want to make sure that the velocity is zero across the surface. To be clear, individual points on each panel are used to evaluate the element (vortex) flow field –it needs its own origin, or coordinate system, to write an equation for the flow generated by this element. We also only pick a point on the panel to check to make sure that the net sum of contributions from all elements results in zero flow across the surface. The fact that we only satisfy the condition at one point on each panel will be satisfactory if the panels are made to be reasonably small. These points are called "collocation points" on each panel. In the end each panel will have coordinates that define its location on the surface, coordinates for the element location on the surface and coordinates for the collocation point.

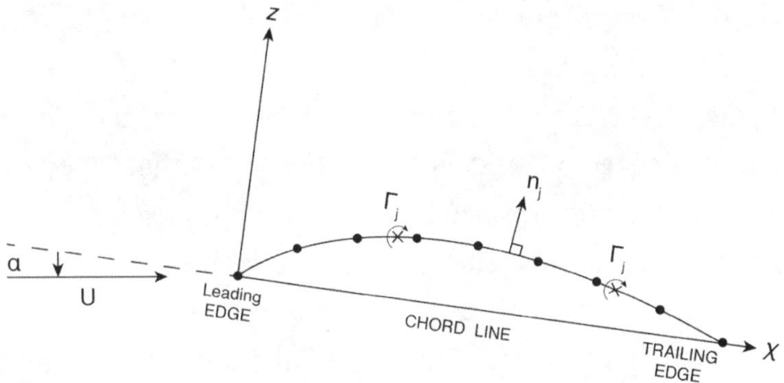

Fig. 6.2 Geometry of panels over a two dimensional wing showing individual vortices, Γ_j associated with each panel; n is the outward normal for each panel; the wing is at an angle of attack to the freestream of α.

The method of solution for the force on the object is the determination of the magnitudes or strengths of the elements on each panel. Once we have this distribution of strengths we can calculate the total lift on the surface that results from all of these elements. Recall that the lift experienced by a surface is really the component of the pressure at the surface integrated over the surface area in a direction normal to the approach velocity vector of the flow – it is not necessarily vertical, but normal to the freestream velocity.

For illustration we are going to use flat plate panels with vortex elements, one per panel. To get an idea of how to define the vortex origins on each panel we can examine flow over a single flat plate at some angle of attack to the freestream velocity vector, α.

First we define the "center of pressure", x_{cp}. This is the location on the surface where the resultant lift force caused by the distributed load on the surface acts such that there is no net moment on the surface (see F.M. White, Chapt. 8, for some details on this). In general, if the moment on the surface about the leading edge is Mo, and L is the lift force then using x as the coordinate measured along the flat surface from the leading edge then:

$$x_{cp} = -Mo/L.$$

Consider now flow over the same flat surface, the lift we have seen is $L = \rho U \Gamma$ note that we are going to define Γ as positive clockwise – this is opposite to what we have done previously, but it helps get rid of some negative signs, this is not necessary, but convenient. To find out the value of x_{cp} we use the theoretical evaluation of the lift force based on "thin airfoil theory". The interested reader can refer to F.M. White Chapt. 8 for details on this. First, assuming a two dimensional flow, we assume that there is some continuous distribution of vorticity over the surface (in contrast to discrete vortex elements), such that there is a local circulation per unit length of surface, $\gamma(x)$. The surface integral of this distribution results in the net total circulation: $\Gamma = \int_0^c \gamma dx$ where "c" is the length of the flat surface, which we call the "chord".

The Kutta condition is imposed where we want the flow to leave the surface flowing parallel with the back (trailing) edge. If this is a surface like an airfoil we want both the top and bottom flows to leave smoothly from the trailing edge. For this to happen then we don't expect there to be a pressure difference between the top and bottom of the surface at the trailing edge, which would cause the flow to deflect up or down. If there is no velocity or pressure difference across the flow at the trailing edge the local circulation at the trailing edge, $(x = c)$, must be zero. So now we have a boundary condition for the integration of $\gamma(x)$ to get the total circulation Γ, that is $\gamma = 0$ at $x = c$. The needed distribution has been figured out for a single flat surface at an angle of attack of α, the angle between the oncoming flow and the surface. The function that works is: $\gamma = 2U \sin\alpha \left(\frac{c}{x} - 1\right)^{1/2}$. The integral of this expression for $\gamma(x)$ yields the value of Γ. Once the circulation is known then the lift, L, can be found. Using this lift, and also using the above definition of x_{cp}, it turns out that the "center of pressure occurs at $x_{cp} = c/4$. So by assuming the lift force occurs at x_{cp} then there is no moment caused by the distributed pressure force.

The lift force, L, as calculated as stated above, can be made nondimensional by divided by $1/2\rho U^2 b$, where b is the span of the surface (in and out of the page). The result is an expression for the lift coefficient:

$$C_L = 2\pi \sin\alpha \tag{6.3}$$

This result is the thin airfoil theory result for the lift on a surface at angle of attack, α. So at this point, we take this result for a flat surface and place vortex elements at the center of pressure for each flat element, which will be at a point ¼ of the distance along the panel from the beginning edge of the panel. Although this is not absolutely required for the panel method it is a convenient choice. As panel size is reduced smaller and smaller the impact of this choice becomes small.

The next step is to find the location where we want to impose the condition of zero velocity crossing the surface for each panel. For this we are using a coordinate system for the panel to be (x, z) where x is, again, along the flat surface from the panel leading edge and z is normal (upward) from the surface. To find these collocation points, we note that the the velocity set up by a vortex element at location r from the origin is $v_\theta = -\Gamma/2\pi r$. Writing this in Cartesian coordinates where at the panel surface $v_\theta = w$ (where w is the z component of velocity) is:

$$w = -\frac{\Gamma}{2\pi}\frac{(x - x_o)}{(z - z_o)^2 + (x - x_o)^2} \tag{6.4}$$

where (x_o, z_o) is taken as the center of pressure (where the vortex element is located). Also, since the collocation point is on the surface then both z and z_o are zero.

To get the total normal component velocity at the panel surface we add together w being the velocity normal to the surface induced by the vortex, with the component of the freestream flow normal to the surface which is at angle α to the surface, $U \sin \alpha$, and set this to zero:

$$w + U\ sin\ \alpha = 0 \tag{6.5}$$

We next insert Eqn. (6.4) for w, with $z_o = 0$ and $x_0 = c/4$. And we define the location at which we evaluate the zero normal velocity component as the collocation point as: $x = kc$. Finally we solve for "k" to be $k = 3/4$. This says that the x location along any given panel where the boundary condition is to be satisfied is at $x = 3\ c/4$. Recall that the vortex element is located at $x_o = c/4$.

In summary, once the panels are set up on the surface, each has its own angle of attack, α_ρ, depending on the local orientation of the surface. Each panel has a vortex element located at its own center of pressure and the normal velocity must be equal to zero at $x = 3\ c/4$ measured from the front of each panel. An equation can be written for the zero velocity condition at the collocation point of a panel by taking the sum of all of the contributions from all of the panels, each with a different vortex strength Γ_ρ, and setting this sum to zero. This will result in N equations (one for each panel) with N unknowns for Γ_ρ for each panel. We can then solve for all Γ_ρ using this set of equations.

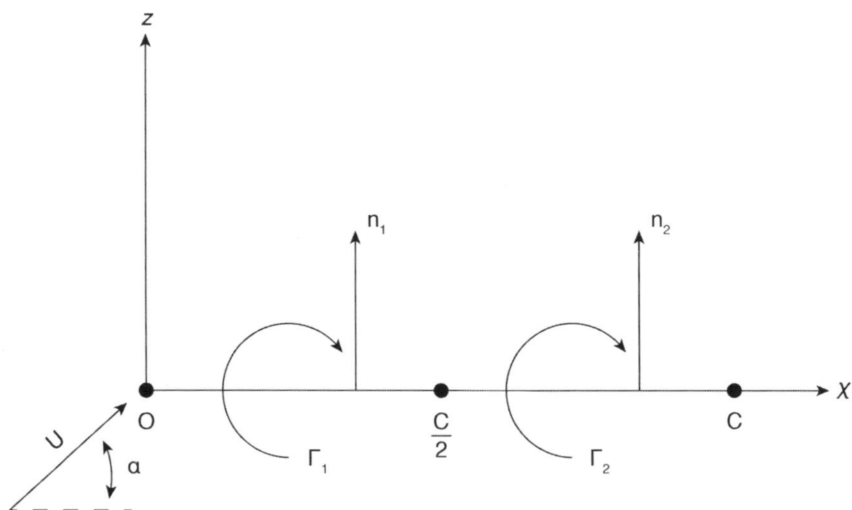

Fig. 6.3: A simple example of two panels along a flat surface of length c *and angle of attack of* α.

Now we can set this up for a simple example to help pull all of this together, see Fig. 6.3. We use a thin flat surface representing a flat airfoil at an angle of attack of α, with a uniform approaching flow of U. We use two panels for illustration. Note that the entire length of the two panels together is "c", NOT the length of each panel. Also, x and z, used previously for a single panel, are along and normal to a line from the start to the end of the surface, respectively. This provides a coordinate system for the "system" of panels. For this simple case the angle of attack for both panels are the same, but we allow individual panel vortex strengths, Γ_1 and Γ_2.

In determining the vortex location for each panel, using the overall coordinate system we get (c/8,0) and (5c/8,0) for panel 1 and panel 2, respectively. The collocation points are (3c/8,0) and (7c/8,0), respectively.

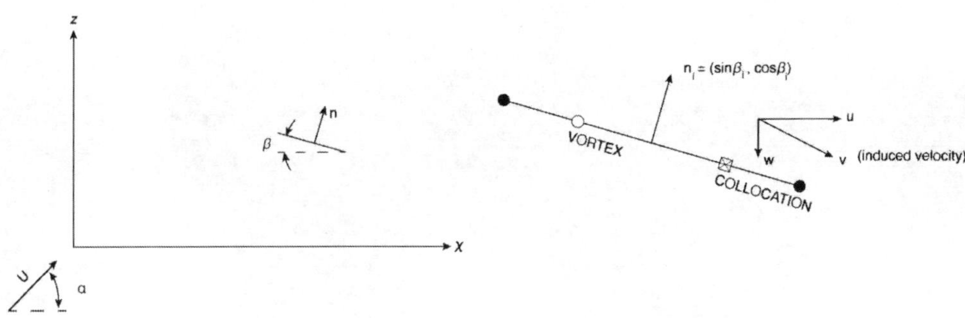

Fig. 6.4: Illustration of the coordinate system using the chord as the x axis; this is the coordinate system typically used to define a_{ij}.

Note that the each outward normal, n_i, points in the same direction for this case since both panels have the same orientation, but in general each panel could be at some angle β_i. Fig. 6.4 shows the general definition of the panel angle to the chord, the latter being along the x axis. The outward normal to the panel is given as n_i. For the general case shown in Fig. 6.4 the outward normal is at an angle β_i to the z axis.

The general equation for the boundary condition is Eqn. (6.5), which we write slightly differently as:

$$v_i \cdot n_i = -U \cdot n_i \tag{6.6}$$

Where v_i is the induced velocity by the set of vorticies evaluated at each panel j and n_i is the panel outward normal. The left hand side is the dot product of the induced velocity with the outward normal. The right hand side is the component of the freestream velocity aligned with the outward normal. These two balance each other to yield zero velocity crossing the panel. The induced flow from all of the panel elements needs to be added together to get the value of v_i in the above equation.

The general form for the velocity, which has components identified as u, w,at any point in the flow with a vortex element located at (x_o, z_o) is obtained from above and written here as:

$$u = \frac{\Gamma}{2\pi} \frac{(z - z_o)}{(x - x_o)^2 + (z - z_o)^2} \tag{6.7}$$

$$w = \frac{-\Gamma}{2\pi} \frac{(x - x_o)}{(x - x_o)^2 + (z - z_o)^2} \tag{6.8}$$

where we have transformed v_θ into the Cartesian components u, w.

Writing this as a matrix we obtain a general expression for the induced velocity anywhere in the flow (x, z):

$$\begin{pmatrix} u \\ w \end{pmatrix} = \frac{\Gamma}{2\pi r^2} \begin{pmatrix} 01 \\ -10 \end{pmatrix} \begin{pmatrix} x - x_o \\ z - z_o \end{pmatrix} \tag{6.9}$$

Note here that $x - x_o$ is the distance along the x axis from a vortex element to the point of interest where the velocity is being evaluated. Similarly for $z - z_o$. Since the values for Γ are not known, it is convenient to write this set of two equations for a "unit value" of Γ, that is $\Gamma = 1$. Keep in mind that the flow caused by each vortex when evaluated on a surface results only in a component normal to the surface. We also are going to use values of x_o and x based on the location of the vortices and collocation points, respectively.

Consequently, for our two panel example, where each panel is along the chord line, or the x axis, we can write the velocity components given below, however here these velocities assume $\Gamma = 1$ in the above equations. The notation used below is that the first subscript, i, represents the collocation point of interest at some panel, and the second subscript, j, identifies the vortex element at some panel that induces the velocity at the i collocation point. That is to say these subscripts do NOT represent components of vectors in space in the conventional manner, and here u and w are the x and z component of the velocity at the surface.

$$(u_{11}, w_{11}) = \left(0, -\frac{1}{\frac{2\pi c}{4}}\right) = (0, -\frac{2}{\pi c}) \tag{6.10}$$

$$(u_{21}, w_{21}) = \left(0, -\frac{1}{\frac{2\pi 3c}{4}}\right) = (0, -\frac{2}{3\pi c}) \tag{6.11}$$

$$(u_{12}, w_{12}) = \left(0, \frac{2}{\pi c}\right) \tag{6.12}$$

$$(u_{22}, w_{22}) = \left(0, -\frac{2}{\pi c}\right) \tag{6.13}$$

For instance, u_{12} is the x component of velocity at the collocation point of panel 1 caused by the vortex element on panel 2. Also, w_{12} is the z component of velocity at collacation point 1 caused by the vortex element on panel 2.

We can define, again for $\Gamma_\rho = 1$, for each panel with vortex j, the following matrix a_{ij}:

$$a_{ij} = \boldsymbol{v}_{ij} \cdot \boldsymbol{n} \tag{6.14}$$

This represents the component of the induced velocity in the outward normal direction for a given panel $\Gamma = 1$. We need to be careful here with this notation, v_{ij} in the equation above is the velocity vector caused by the vorticies with unit value circulation with the subscripts: i represents the collocation point and j the vortex location inducing that flow. Also, the dot product with n (the outward normal for each panel), gives the projection of v normal to the surface. For example for the above example n is only in the z direction.

$$a_{11} = \left(0, -\frac{2}{\pi c}\right) \cdot (0.1) = -\frac{2}{\pi c} \tag{6.15}$$

so the elements a_{ij} are quantities representing a measure of the velocity contribution normal to each panel from each vortex j at each collocation point i for a $\Gamma_j = 1$ at each panel.

Once we have identified the components a_{ij}, which, as shown above for the two panel example, are determined only by geometry, it is possible to use the impervious boundary condition to find each of the values for Γ_j. We have a general equation for our two element surface for panels $i = 1, 2$ that says the sum of the outward normal velocity at each panel with the freestream component normal to that panel must be zero. The outward normal velocity due to summation of the induced velocity over both panels is for some unknown distribution of circulations, Γ_j:

$$\sum_{j=1}^{2} a_{ij}\, \Gamma_j \tag{6.16}$$

So this expression is balanced by the contribution from the free stream velocity: $-U \cdot n$ (this is the component of U normal to the surface).

The final system of equations become:

$$\sum_{j=1}^{2} a_{ij}\Gamma_j = -(U_x, U_z) \cdot (n_x,\, n_z) \tag{6.17}$$

Here the right hand side value of U is represented as the vector (U_x, U_z) indicating that U is a vector with x and z components that depends on the angle of attack given by the chord line. The x, z components of the outward normal will depend on the angle β for each panel. The geometry of all of the panels fully determines a_{ij}, as shown above in the simple two panel example. Consequently, once the geometry is known the values of a_{ij} and $(n_x, \ n_z)$ are all determined. Then knowing U the set of Eqns. (6.17) provides the means to find all of the values of Γ_j, where the index represents each of the panel vortices. In the above two panel example there are two values of Γ.

From the solution of Γ_j we can find the lift on each panel, $\Delta L_j = \rho U \Gamma_j$ and then the total lift as the sum over all panels:

$$\sum_{j=1}^{2} \Delta L_j = C_L \left(1/2 \rho U^2 c\right) \tag{6.18}$$

Recall that the lift is normal to the direction of the freestream velocity, so the lift of each panel, even though the panels may all be at different angles of attack, are all in the same direction.

It is also possible at this point to calculate the velocity tangent to the surface at each panel. That is to say, instead of finding the component normal to the surface to satisfy the impermeable boundary condition find the component tangent to the panel, for each panel and combine this with the component of U tangent to the panel. The induced velocity tangent to the panel can be found using the dot product of the vortex induced velocity with the unit vector tangent to the panel. The tangent unit vector is given by its x and z components as $cos\beta_i$, $sin\beta_i$, respectively with β_i being the angle of the panel relative to the chord line which is the x axis, show in Fig 6.4. The freestream contribution is found from the value of U. This combined velocity yields the velocity vector of the flow at each panel and can be used in the Bernoulli equation to find the pressure at each panel. The Bernoulli equation is applied from a known upstream condition (known pressure and freestream velocity) to each panel in order to calculate the panel pressure.

Thin Airfoil Theory Overview

This is an overview of what is known as "Thin Airfoil Theory" used to develop some major results used to analysis flow over airfoils in general, and provide insight into the nature and trends of lift forces generated by flow over airfoils. Airfoils in general come in different geometries and shapes that induce flow perturbations on an approaching flow. These perturbations to the velocity field result in pressure changes that cause a net force on the foil. The force is typically divided into the component aligned with the oncoming flow direction of the fluid (drag force) and the force normal to the oncoming flow direction, (lift force). Viscous effects are not included, but the interested reader may explore some of the options of how certain viscous effects are included by researching on the internet (for instance J.E. Yates, 1980, "Viscous Thin Airfoil Theory", Aeronautical Research

Associates of Princeton, Report 413, as well as several more recent developments). In this short description we are limited to mostly two-dimensional steady flow, although the theory can be extended. Thin airfoil theory can also be extended to unsteady flows such as vibrating or oscillating foils or rotating airfoils (see J. Katz and A. Plotkin, 2010, Low Speed Aerodynamics, Cambridge University Press,).

The designation of a "thin" foil provides several mathematical conveniences, which includes the use of "small perturbations" to the freestream flow. This allows some simplification of terms in the governing equations. These mathematical details are not discussed here, but are available in more advanced book on aerodynamics (see J. Katz and A. Plotkin, Low Speed Aerodynamics) The geometric terms used to define the geometry of an airfoil are shown in Fig. 6.5. The assumption is that the foil thickness (its maximum value) is much less than the chord length, c, of the foil. The chord length is measured from the leading edge to the trailing edge (in a straight line called the chord line) of the foil. The leading edge is at the front tip and trailing edge is at the rear tip. This can be seen in the figure below which shows the chord length, thickness and also the "camber line" for a given cross section of a wing. The camber line is a line extending from the leading edge to trailing edge that defines the mid-point between the top and bottom of the foil and its effects are discussed later. The degree to which this is curved (rather than falling directly on top of the chord line) defines the "camber".

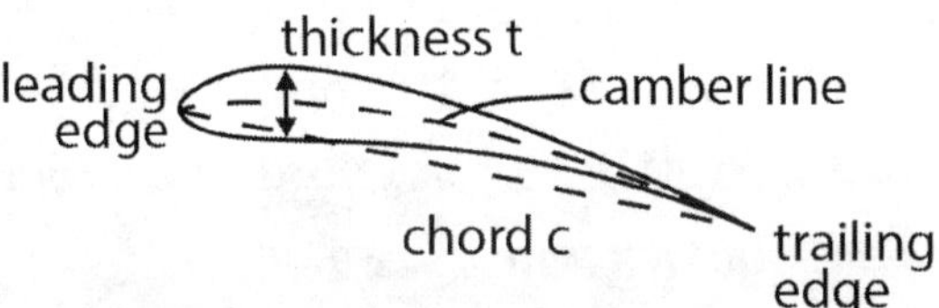

Fig. 6.5 Illustration of the basic elements used to define an airfoil, the chord line and camber line and foil thickness; in thin airfoil theory the foil is replaced by the camber line, ignoring the thickness effects.

The general flow situation needed to generate a lift force involves asymmetry between the flow above and below the foil. As flow goes over and under the foil (from the left in the figure), due to asymmetry the flow over the top will be different from that on the bottom. Asymmetry can be set up in two ways, one by foil shape and the other by rotating the foil relative to the freestream flow direction. If the chord line is curved upward providing camber (see figure) then the flow path along the top and bottom will be different and the velocities most likely will differ. The other way to create asymmetry is to tilt the foil relative to the flow of the fluid, U, this tilt is called the angle of attack, α. Specifically, the angle α is between the chord line and the freestream velocity. The velocity difference between the top and bottom flow will imply that the pressure distribution along the top and bottom are also different from each other. A net force results acting on the foil by this pressure difference. Integrating the pressure difference along the entire foil yields the net force.

In thin airfoil theory, typically an inviscid flow is assumed (no consideration of frictional force between the flow and the foil). However, viscous forces occur at the foil surface and are assumed to result in vorticity perturbations along the surface. The foil thickness is taken as thin so the top and bottom surfaces are taken together as a single surface with flow over the top and bottom along the camber line. The flow is analyzed using a distribution of local vorticity placed along the camber line. This is called a vortex sheet (in that it extends into the span of the foil) and is shown in Fig. 6.6. as a series of small vortices distributed along the foil from leading edge to trailing edge. We denote this distribution by $\gamma(x)$ which represents the local circulation per length dx. Recall circulation has units of velocity times distance (m2/s), so $\gamma(x)$ has units of (m/s). Note that $\gamma(x)$ is not a constant along the foil. For this analysis we assume that circulation is positive if rotation is clockwise. This is opposite to the more conventional sign for circulation previously used, but this change is convenient for airfoils since this designation allows for the generation of upward lift for positive circulation.

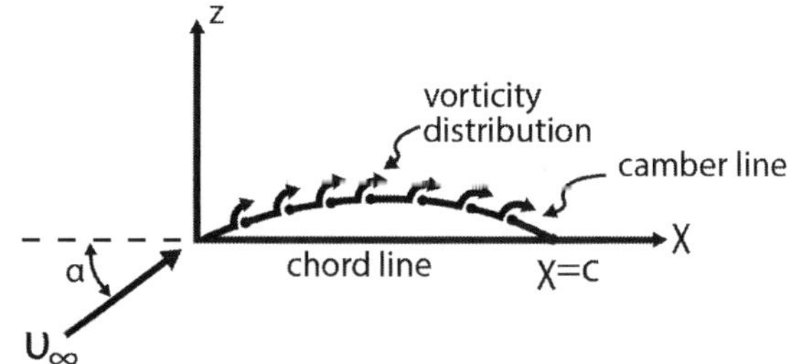

Fig. 6.6 Geometry used to define the flow configuration of a thin airfoil where [latex]\alpha[/latex] is the angle of attack (angle between the approaching free stream and the chord line); a distribution of vorticity is assumed along the airfoil shown here as discrete vortex elements but in thin airfoil theory is assumed to be a continuous distribution.

General Formulation of Coefficient of Lift

The flow over the foil in reality is viscous flow and it forms a viscous layer of flow (a boundary layer) along the foil. Within this thin layer is vorticity based on the local magnitude of the velocity derivative, $\partial u/\partial y$, where y is the coordinate normal to the foil surface. As was mentioned, this distribution of vorticity gives rise to the local circulation distribution $\gamma(x)$. Also, the integration of $\gamma(x)$ along the entire foil results in a value of total circulation for the foil, Γ, such that:

$$\Gamma = \int_0^c \gamma(x)\, dx$$

Because the foil is a solid surface with no flow across it. Then we know that the circulation set up in the boundary layer is not able to cause flow into the foil, rather by superposition, the sum of the flow induced by the local vortex flow, dv_v, determined by the magnitude of γ(x), must be balanced by the freestream flow component normal to the foil surface (which is at angle α to the freestream if we neglect the small camber for now). Since these two velocities must balance yielding no flow across the foil, then the vortex induced velocity can be written as:

$$dv_v = U \sin \alpha$$

Note that the vortex induced velocity is given by $\gamma(x)dx = 2\pi r dv_v$, where r is the distance from the center of the vortex to some point along the foil surface. To get the entire induced velocity integration is required along the entire foil chord length, c. This can be expressed, for a flat airfoil at an angle of attack of α as:

$$\int_0^c dv_v = \int_0^c \frac{\gamma(x)}{2\pi r} dx$$

The above integral would need to be evaluated along the foil containing a line distribution of vortices given by γ(x) to find the induced velocity at any point on the foil. To do this the variable position "r" needs to be expressed in terms of the coordinate x, measured along the chord line. If we select a position on the foil, x_o, that designated a given location of a vortex sheet element then $r = (x_o - x)$. Using the above expression for dv_v in terms of $U \sin \alpha$ we obtain an integral equation for the distribution of local circulation:

$$U \sin \alpha = \int_0^c \frac{\gamma(x)}{2\pi(x_o - x)} dx \tag{6.19}$$

The above condition needs to be satisfied for the given distribution $\gamma(x)$ along the foil. However, we need a boundary condition at $x = c$ that specifies $\gamma(c)$. This is done by introducing what is known as the Kutta Condition. This is based on the observed flow condition at the trailing edge of the foil, assuming the geometry is such that the flow from the top and bottom sides meet at a point (the trailing edge) and that flow exits the foil "smoothly" from the top and bottom surfaces such that there is no vorticity as the flow leaves the trailing edge. Physically this would occur if there is no roll-up of the flow say from the bottom to the top surface. If the velocities on top and bottom at the trailing edge are identical, then the pressures are identical by the Bernoulli equation, and the flow has no tendency to swirl. So, this condition is expressed as $\gamma(x = c) = 0$. It can then be shown that the integration in Eqn. (6.19) is found to satisfy the Kutta Condition for the following distribution of the vortex sheet:

$$\gamma(x) = 2U \sin\alpha(c/x - 1) \tag{6.20}$$

(See the extra comments on the Kutta condition below)

With this distribution of $\gamma(x)$ the total circulation, Γ, can be determined. Note that this value of Γ depends on the angle of attack, α, and the free stream velocity, U. We can now use the previously given result that the lift force, L , is dependent on the circulation that was shown for a rotating cylinder in a freestream, $L = \rho U \Gamma$. This is further discussed below and is found to be valid for airfoils in irrotational flow.

The integration of $\gamma(x))$ given in Eqn. (6.19) results in the following Lift expression for an airfoil:

$$L = 2\pi \sin\alpha \rho U^2 c$$

Or by defining the nondimensional lift coefficient as the lift force divided by $1/2\rho U^2$:

$$C_L = 2\pi \sin\alpha \tag{6.21}$$

This surprisingly simple expression is a major finding of the thin air foil theory in the description of the lift force as determined by the foil angle of attack. This is based on a two-dimensional steady flow (or infinitely wide foil) with very little camber with viscous effects ignored. Some additional features and effects will be added to this equation later. Figure 6.7, shows the lift coefficient versus angle of attack. At large angles of attack the results of eqn. (6.21) are not valid. This is because the angle is so large that flow over the leading edge cannot negotiate the needed turn to stay "attached" to the foil surface. Rather the flow "separates" from following the surface contour. This results in a recirculating flow region between the streamlines of flow and the body surface. This can be seen in Fig. 4.5 illustrating flow going over the leading edge of a thin airfoil at a fairly large angle of attack.

The primary limitation on the application of thin airfoil theory is the assumption of inviscid flow (which is actually has been shown to be a fairly valid approximation since all of the friction induced vorticity is accounted for with the vortex sheet). Also, the foil must be thin (maximum thickness on the order of ten percent of the chord length). This is not necessarily that restrictive since even airfoils with thicknesses of over 10% of the chord follow thin airfoil theory fairly well. With some modifications to the basic lift result of Eqn. (6.19) thickness effects, relatively large camber and three-dimensional effects can be accounted for, as given below. A generic plot of the lift coefficient, C_L, versus angle of attack, a, is given in Fig. 6.7, as well as results for a specific airfoil, the NACA 0012 compared with Eqn. (6.21). Very good agreement between data and theory are shown up to an angle of attack of about 16°.

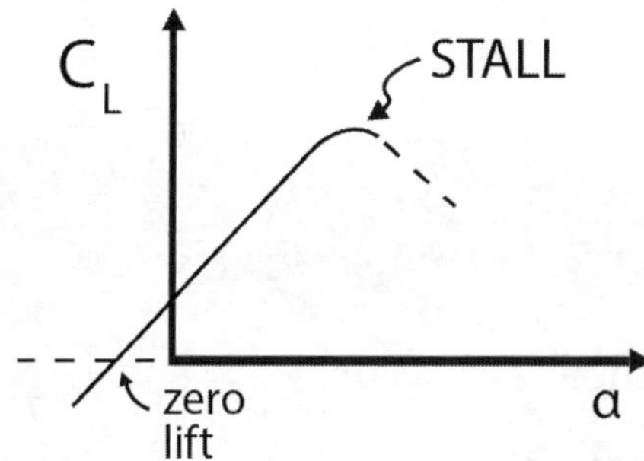

Fig. 6.7. Illustration of a typical Lift coefficient versus angle of attack curve, left, and comparison with Eq1n. (6.19) and data for a NACA 0012 air foil.

The Kutta Condition

(see https://en.wikipedia.org/wiki/Kutta_condition)

As mentioned above a boundary condition was needed in order to complete the thin airfoil theory to determine ultimately the lift force on an airfoil. This is known as the Kutta condition and is applied to airfoils when describing the external flow around the foil to provide an inviscid condition that is linked to the viscous flow effect at sharp corners on an object. First consider inviscid flow over a cylinder (two-dimensional flow). In such a flow there is no separation and there is a stagnation point on the upstream side of the cylinder and a similar stagnation point on the downstream side of the cylinder. In fact, by observing streamlines it is not apparent which is the upstream and downstream side of the cylinder since the streamlines, and the flow in general is symmetric. This symmetric flow results in a symmetric pressure field around the cylinder such that the drag and lift forces are identically zero. If the object is changed to be an oval the flow again has two stagnation points, and when the oval major axis does not align with the flow direction there is also two stagnation points that do not align with the major axis. Now consider the case of an airfoil, with a rounded leading edge and a very sharp trailing edge. If the airfoil has a slight positive angle of attack then the upstream stagnation point occurs on the lower surface of the airfoil. The downstream stagnation point will want to move to the topside of the airfoil surface as for an oval shape. But in doing this the flow must go around the sharp trailing edge which has a near zero radius of curvature. Consequently, the velocity required to go around this corner tends towards infinity. In reality viscous effects acting on this high velocity region prevent the occurrence of this very large velocity. The Kutta condition accounts for this by forcing the velocity on the upper and lower surface to be identical at the tip of the sharp trailing edge. This results in a zero pressure gradient across the flow at the trailing edge as well. If the airfoil were to start from rest and suddenly accelerate within a fluid at rest then the flow would role-up behind the trailing edge forming a vortex whose strength depends on the acceleration. This shed vortex leaves the airfoil as the airfoil moves, this is known as the "starting vortex".

By Kelvin's theorem for an inviscid fluid there cannot be any change in total vorticity within a fluid region so the airfoil must carry with it a "bound" vorticity equal and opposite in sign to the starting vorticity. Since circulation is the spatial integral of vorticity then it can be said that the airfoil has circulation, which can be shown to be proportional to the lift force generated on the airfoil by the Kutta-Joukowski theorem. What is left is a stagnation point at the trailing edge of the airfoil as the flow from the top and bottom surfaces come together, similar to the flow from the top and bottom of the original cylinder discussed earlier. The fluid leaving the top and bottom surfaces flows parallel with each other with equal magnitude. This is a convenient physical flow condition that is often used in many flow conditions to provide mathematical closure.

Circulatory Lift: A Simple Method to Define Lift for an Airfoil

In this section we present an alternative approach to determining the lift for an airfoil. Consider a two-dimensional steady flow (a very wide airfoil) with the goal of determining the lift force versus angle of attack. There are much more rigorous methods to do this, involving complex variable representation of the flow. However, this simple treatment is rather straight forward and yields the same result in a more intuitive manner. The goal here is to illustrate from a flow physics point of view how the circulation is related to the lift force, and without circulation the lift force vanishes.

The no-slip boundary condition at the surface of the airfoil generates vorticity along the surface of the wing. For a perfectly symmetric wing that is oriented along the incoming flow direction the streamlines above and below the foil are identical. One can also imagine that the vorticity distribution on the top of the wing generated by surface friction is identical in magnitude but opposite in sign on the bottom of the wing (the rotation direction is opposite). Consequently, the sum total of vorticity around the wing becomes zero for this situation. However, tilting the wing at a given angle of attack relative to an in-coming flow creates an asymmetry of flow. See Fig. 6.8 below. Notice in the figure that the flow leaves the trailing edge of the wing with a net downward flow direction either due to the camber or a tilting of the chord line relative to the freestream flow direction. In fact, in a vector sense the downward momentum of the flow has to have been generated by some downward force acting on the fluid. Selecting a rectangular control volume surrounding the wing and using this to integrate the velocity field along the surface of the control volume area as shown determines the magnitude of the circulation as:

$$\Gamma = \oint V \bullet ds = \iint_A \omega \bullet dA$$

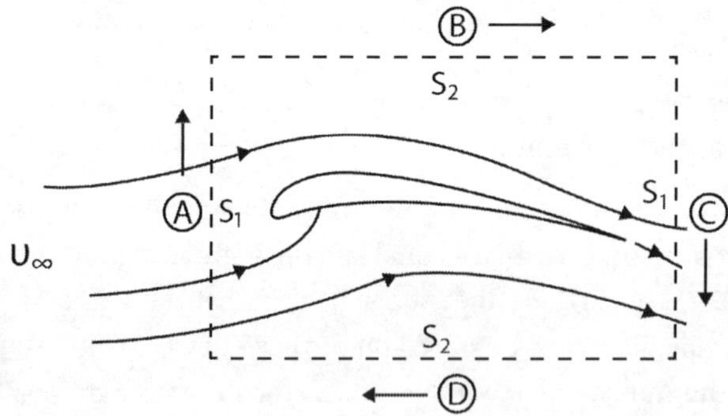

Fig. 6.8. Illustration of the streamlines associated with flow over a cambered airfoil producing asymmetric flow with a freestream velocity of [latex]U_\infty[/latex]; a control volume analysis is used with sides A, B, C and D.

Where the line integral of the velocity is around the boundary of the area associated with the area of the vorticity integral. The contributions to the line integral for the different sides of the area are:

A: = 0 (there is no velocity component along this length (dot product of V and ds is zero)

B: ~$U_\infty S_2$ (the boundary is far from the wing surface so the velocity is close to U_∞)

C: ~$Ws1S_1$ (W is the downward velocity component of the flow coming off of the trailing -edge which is assumed to be parallel with the top of the airfoil)

D: ~$-U_\infty S_2$ (negative since the velocity is in the opposite direction as the integration)

By summing these components to obtain the total integration becomes $\Gamma = WS_1$. This shows that the circulation is equal to the amount of downflow generated as the flow passes the airfoil. The lift force on the wing is equal and opposite to the force on the fluid. Using the momentum equation to determine the relationship between the force on the fluid and the change of momentum of the fluid results in (where forces are primed to represent force per span into the page):

$$F' = L' = m^{\cdot} W = (\rho S_1 U_\infty)W = \rho U_\infty)\Gamma$$

This illustrates that any object generating a downflow velocity in the control region will result in a net lift force on the object balancing the change of momentum of the fluid in the vertical direction. Also, this force is seen to be directly proportional to the circulation set up around the object. The assumption here is of irrotational flow outside of the viscous boundary layer so that there are no other sources of vorticity inside the selected control volume.

A slightly modified version of this can be used to determine the lift coefficient as well. Imaging a circle drawn around the wing with its center at the center of the wing (radius of $c/2$, where c is the chord length). With this radius the circle will pass through the leading edge and trailing edge. Performing the line integral of the velocity component aligned with the circumference of the circle (dotting the velocity with ds) around the entire circle and using the notation of "W" being the net velocity component downward around the circle results in:

$$\Gamma \oint V \bullet ds = W \oint ds = W 2\pi \left(\frac{c}{2}\right) = \pi c W$$

The lift force per span is then

$$L' = \rho s U_\infty (\pi c W)$$

The lift coefficient is the lift force per area (sc) divided by \frac{1}{2}{{\rho\ U}_\infty}^2: C_L=\frac{L\prime}{1/2\rho{U_\infty}^2}=\frac{2\pi W}{U_\infty}

The ratio of the downwash velocity to the freestream velocity can be estimated as sin a, with a being the angle of attack, which results in:

$C_L = 2\pi \sin \alpha$ ~ $2\pi\alpha$ (for small angle a)

This is the same result obtained using the more rigorous thin airfoil theory which helps to provide some physical insight into the role of the airfoil in defecting flow to obtain a net lift force.

Other Considerations of Forces

As stated previously the lift force is one component of the total force, the other component being the drag force. The drag force, D, can be expressed as a drag coefficient, $C_D = D/(1/2\rho U^2 sc)$, similar to the lift coefficient. Although the drag force varies with angle of attack, unlike the lift coefficient it is fairly constant for a wide range of angle of attack prior to separation as shown in the figure below, except at fairly large angle of attack approaching the stall limit. Near zero angle of attack for a symmetric airfoil this drag is dominated by frictional forces acting on the foil surface. Another representation for the lift and drag coefficient is a "polar" plot which shows how the lift coefficient changes with drag coefficient. An example is shown below where the tangent line provides the minimum lift to drag ratio.

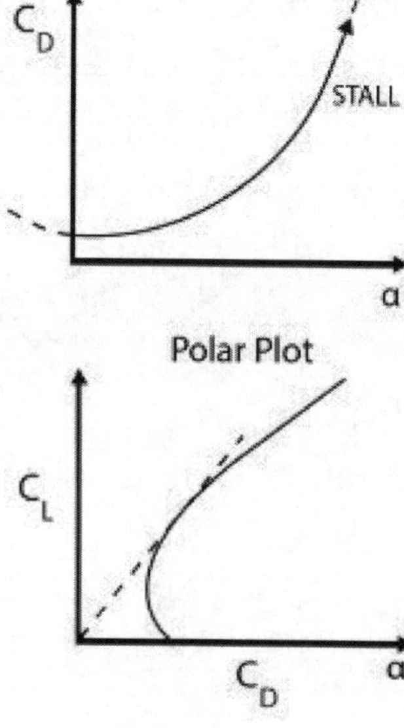

Figure 6.9. Illustration of a typical airfoil drag coefficient, [latex]C_D[/latex], versus angle of attack, top, and lift coefficient versus drag coefficient, known as a polar plot.

As stated previously there may be effects on the lift and drag forces caused by finite span (three-dimensional) wings, thickness and camber that would need to be taken into account. Finite span wings are obviously a reality and the fact that the ends of the wing (wing tips) causes flow conditions in the tip region to change. Specifically, the high pressure that exists below the wing meets with the low pressure on the top of the wing creating the tendency for a tip vortex to form. The resultant flow is a swirling flow directed from the bottom to the top of the wing. This appears as a vortex flow whose axis of rotation aligns closely with the direction of the freestream velocity. As the wing moves through the fluid the vortex axis extends out behind the wing. In fact, this phenomena for airplanes flying at sufficiently high altitudes the vortex motion results in a very low pressure at the center of the tip vorticities. The drop in pressure can cause water vapor to condense resulting in vapor trails or contrails forming at the wing tips that can be seen on the ground as a plan passes overhead. This phenomenon does have some negative effects on the performance of the wing.

The span of a wing determines how significant the tip vortex is on overall forces. A wing can be designated based on its span to chord length ratio (or aspect ratio). The term wing span of an aircraft typically includes the distance from tip to tip of the wings on an airplane – or twice the span of each wing. For a given wing however, the span is from the root to the tip. Denoting span by "s" and chord by "c" the aspect ratio is then s/c. The swirling flow that results near the tip has flow coming up from the bottom, around the tip and then down onto the top of the wing. This downflow will have a tendency to divert the oncoming flow over the wing downward such that there is effectively a smaller angle of attack as seen by the wing. The axis of the vortex near the tip is characterized by the vortex line shown in the figure below extending backwards from the tip. In addition to this a "bound" vortex occurs around the wing due to the previously discussed vorticity distribution, with an axis of rotation along the wing span.

An induced angle of attack is defined as the amount of change (reduction) of the angle of attack as caused by the top vortex:

$$\alpha_{eff} = \alpha - \alpha_{induced}$$

The strength of $\alpha_{induced}$ is given in terms of the downflow velocity induced by the tip vortex as:

$$\alpha_{induced} = tan^{-1} \frac{W}{U_\infty} \frac{W}{U_\infty}$$

Where the later approximation is due to small angles since $W << U_\infty$. Shown here without derivation, the total circulation accounting for the downwash effect written in terms of α_{eff} is:

$$\Gamma \pi c U_\infty \alpha_{eff}$$

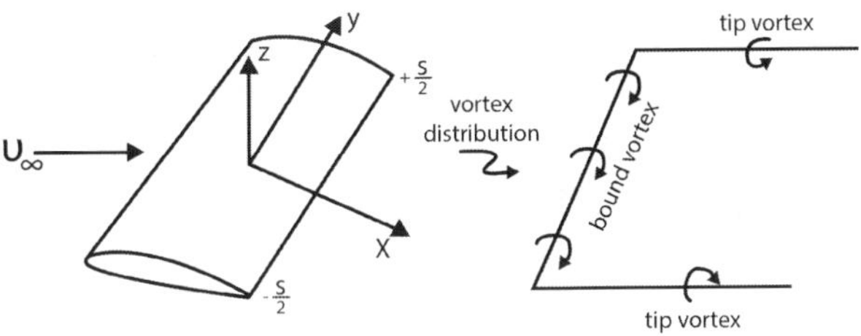

Fig. 6. 10. Illustration of the vortex distribution for a finite length wing; lines represent the vortex axis of rotation.

Using this expression for the circulation to determine the lift after some manipulation the lift coefficient becomes:

$$C_L = \frac{2\pi\alpha}{1 + \frac{2c}{s}}$$

In the limit of large aspect ratio, or c/s becoming small, the lift coefficient reverts to that for an infinite span wing as expected. Also, there is an induced drag that is set up from the downflow which can be determined in terms of the lift coefficient to be:

$$C_{D,\ induced} = \frac{{C_L}^2}{\pi s/c}$$

$$C_D = C_{Do} + C_{D,\ induced}$$

where CD, o is the drag coefficient for an infinite wing.

There is also an effect caused by asymmetry in the foil geometry such as camber. In this case the camber adds to the asymmetry of the foil as shown in Fig. 6.6. The maximum camber along the chord line is designated as "h" and the maximum thickness as "t".

For cambered foils the lift becomes:

$$\Gamma = \pi s U_\infty \left(1 + 0.77\frac{t}{c}\right) \sin\left(\alpha + \beta\right)$$

$$\beta tan^{-1}\left(\frac{2h}{c}\right)$$

As the camber increases, by increasing h/c, and therefor increasing β, the circulation increases. Using this to define the lift coefficient we have the following relationship that accounts for both finite span wings and camber asymmetry:

$$C_L = 2\pi\left(1 + 0.77\frac{t}{c}\right) \sin\left(\alpha + \beta\right) / \left(1 + \frac{2c}{s}\right)$$

The effect of an artificially imposed camber is shown in the figure below by using a flap at the rear of the foil. There can be a significant increase in lift when deploying a flap, and this provides a means to increase lift as needed.

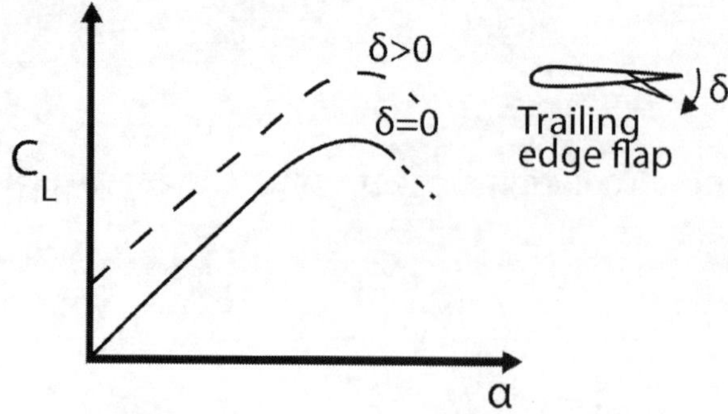

Fig. 6.11. Effect of a trailing edge flap to provide effective camber and increased lift.

Airfoil shapes are often designated by different "codes" or "designations". An early form of this was developed by NACA (National Advisory Committee for Aeronautics, a precursor to NASA). Several such designations have been developed such as the 4, 5 and 6 digit numerical designations as are shown below based on camber, location of maximum camber and thickness. More elaborate designations have been developed over the years.

NACA DIGIT Airfoil Designation

Four Digit: a b c d

a: maximum camber % of chord, c

b: location of maximum camber from leading edge, tenths of c

c & d: maximum thickness, % of c

Example: NACA 2412: max camber is 0.02c; location is 0.4c from leading edge and max thickness is O. l 2c

Five Digit: a b c d e

a: multiple by 2/3 = design lift coefficient in tenths

b & c: divided by 2 = location of maximum camber from leading edge in %c d & e: maximum thickness in % of c.

Example: NACA 2412: design lift coefficient is 0.3; location of max camber is 0.15c, max thickness is O. l 2c

Six Digit: a b c d e f

a & b: series designation

c: location of minimum pressure in tenths of c from leading edge d: design lift in tenths

e & f: maximum thickness in % of c.

VII. Introduction to Viscous Flows

In this section we develop the governing equations for viscous flows resulting in the Navier-Stokes equations. We will simplify the equations for incompressible constant property flows, which are useful for a vast majority of flow situations. We will then show how this seemingly formidable set of equations can be simplified for a number of rather practical flow problems resulting in exact, analytical solutions. We end with the introduction to similarity solutions to the Navier-Stokes equations through an example problem.

Basics of Viscous Forces

The introduction of viscous forces requires a model to obtain a set of conditions on the flow field to express the viscous stress tensor, τ_{ij}, as a function of the local velocity field. This stress tensor consists of the various stresses that can occur on an element of fluid. These forces act at the surface of a fluid element and are caused by a velocity gradient normal to the surface. This velocity gradient can give rise to local friction. In regions of the flow where the velocity gradients go to zero we expect there to be no viscous forces acting. We will show that even though viscous forces may not be zero due to existing velocity gradients their net effect may in fact be zero on a fluid element when taking into account all viscous forces over all the surfaces of an element. Additionally, it is possible to have velocity gradients and no net viscous forces. This will be evident once we present a model equation describing the viscous forces.

By including viscous forces in the flow we need to introduce the no-slip boundary condition, whereby the fluid in contact with a surface takes on the same velocity as the surface — the fluid is not allowed to "slip" along the surface. In doing this a velocity profile is set up across the flow direction. For example in pipe flow the fluid velocity at the pipe wall is the same as the velocity of the pipe, but to have a net flow through the pipe fluid away from the wall is moving differently than the pipe. If the pipe is circular the center of the pipe is furthest from the walls and the fluid in this portion of the pipe is traveling fastest. So for a circular pipe the flow is symmetric circumferentially with a maximum in the center. For pipes with different cross sections we still expect a peak velocity in the region furthest from all walls, since the impact of the wall friction is furthest away. Friction occurs throughout the cross section however, not just at the walls. This internal friction is fluid-fluid friction due to the fact that different "layers" of fluid, or fluid elements are moving at different velocities.

For laminar pipe flows, which most readers have seen in their first course in fluid mechanics, there are some characteristic conditions we can discuss that are relevant to internal viscous flows in general. In the case of steady, fully developed flow we can set a force balance to better understand the required forces acting on the fluid. We will define this characterization of the flow (steady, fully developed) in a later part of this chapter, in reference to the governing equation. Referring to Fig. 7.1 (a) for the illustrated control volume of length dx extending across the area of the round pipe of radius R we see that a force balance for nonaccelerating flow between the force due to pressure and that due to wall friction becomes:

$$\frac{dP}{dx} = -\frac{2\pi R}{\pi R^2}\tau_w = -\frac{2\tau_w}{R} \tag{7.1}$$

This indicates that the pressure decreases along the flow direction proportional to the magnitude of the wall shear stress caused by friction between the fluid and pipe wall. In this equation the values of the wall shear stress is taken as a positive number.

External flows have a similar no-slip condition at a surface, the difference being that as one moves away from the surface any other boundaries are “infinitely” far away in the sense that these other surfaces are not influencing the flow. This is shown in Fig. 7.1 (b) showing that the extent of the friction layer increases along the flow direction. However, there is still wall friction and fluid-fluid friction up to some point far enough from the surface, which depends on the particular location along the surface. We will discuss this in detail when we examine boundary layer flows. Recall that in previous sections we neglected viscous effects for external flows and the results tend to be very accurate, this is because for the flows we looked at other forces that come into play (pressure and body forces) dominate the flow. However, even in these flows viscous forces can be important in that they lead to “flow separation” where streamlines no longer follow the general shape of the object. When this occurs the potential flow solutions tend to break down, even though pressure forces dominate the overall forces. This is because the velocity distribution is greatly altered by the flow separation. This topic of flow separation is beyond the scope of this book but we will discuss it to a limited extent in the next chapter. The reader is referred to “Viscous Fluid Flow” by F.M. White for a discussion of this and some of the many models used in these flows, to predict wall friction effects.

The velocity profiles shown in Fig. 7.1 indicates a variation of velocity from the boundary condition at the surface of no-slip to a maximum at the centerline for pipe flow or reaching the freestream velocity far from the surface in external flows. This implies a distribution of internal viscous stress caused by the local velocity distribution since frictional forces depend on the relative velocity variation, or velocity derivative, within the flow field. In order to determine the magnitude of these forces we will first define the viscous stress tensor.

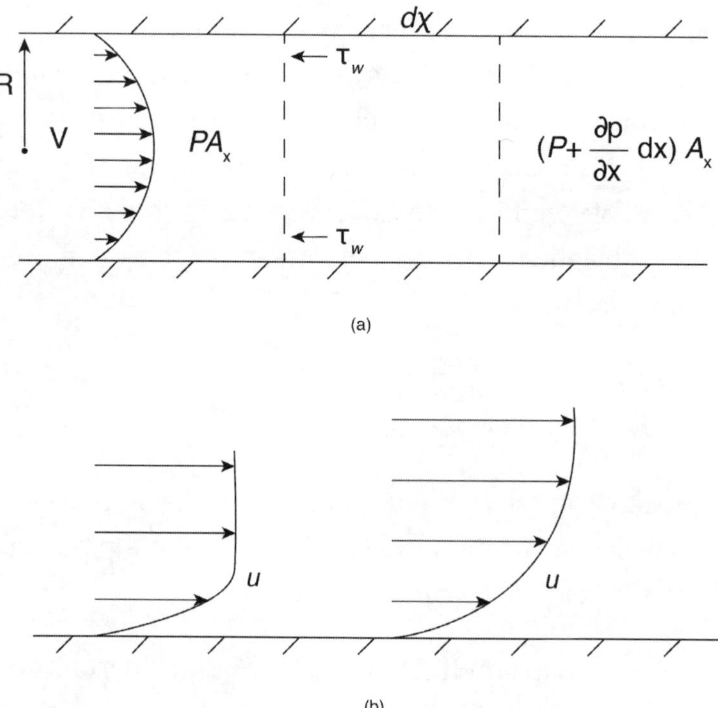

Fig. 7.1 Illustration of velocity profiles (a) steady fully developed pipe flow resulting in a balance of pressure forces and wall friction forces, (b) external flow over a flat surface indicating boundary layer profile.

Viscous Stress

Viscous forces are represented here using the viscous stress tensor, τ_{ij}, that contains nine possible elements in three-dimensional space, three diagonal elements, i=j, and six off diagonal terms. Each of these terms will be given an interpretation relative to the flow. Physically the viscous stress is linked to the velocity gradient that results in a strain rate acting on a fluid element. Recall that fluids can have a continuous deformation with an applied force, in contrast to a solid which may have a given deformation proportional to the applied stress. We anticipate a relationship between the stress and the resultant strain rate occurring in the fluid. So first we define the strain rate that occurs within the fluid.

A small fluid element is shown in Fig. 7.2. First consider Fig. 7.2 (a) showing a "linear strain". Assume there exists an increase of u_1 in the x_1 direction, $\frac{\partial u_1}{\partial x_1} dx_1$, as a result, the surface of the element on the right (at larger x_1) will move faster to the right than the surface on the left (at smaller x_1). Over time Δt the distance between the two surfaces, denoted as "l" which is originally dx_1 will increase by Δl leading to a rate change of separation distance, or a rate of deformation equal to:

$$\frac{1}{l}\left(\frac{\triangle\, l}{\triangle\, t}\right)=\frac{\triangle\, l/l}{\triangle\, t}=\frac{\partial u_1}{\partial x_1} \tag{7.2}$$

Treating $\Delta l/l$ as the strain then Eqn. (7.2) represents the linear strain rate in the x_1 direction caused by the u_1 velocity gradient in the x_1 direction. We can write similar expressions for the x_2 and x_3 directions as $\frac{\partial u_2}{\partial x_2}$ and $\frac{\partial u_3}{\partial x_3}$, respectively. Physically, for positive values of these velocity gradient components in each direction we see that there is a stretching in each direction, there is similar compression if the derivatives are negative. There is a change of volume per time of $\frac{(\triangle\, l/l)_1}{\triangle\, t}(\partial x_1 \partial x_2 \partial x_3)=\frac{\partial u_1}{\partial x_1}(\partial x_1 \partial x_2 \partial x_3)$. Similarly in all 3 directions. Taken together over time there is a change in volume of the fluid element measured by the sum of the total change in each direction. Dividing through by volume $(\partial x_1 \partial x_2 \partial x_3)$ we call this result the "dilation rate" of the fluid. Using tensor notation we can write the dilation rate (using the summation rule) as:

$$\frac{\partial u_i}{\partial x_i}=\dot{e}_{ii} \tag{7.3}$$

(Where summation is implied)

The units of $\dot{e}_{ii}$ are one over time.

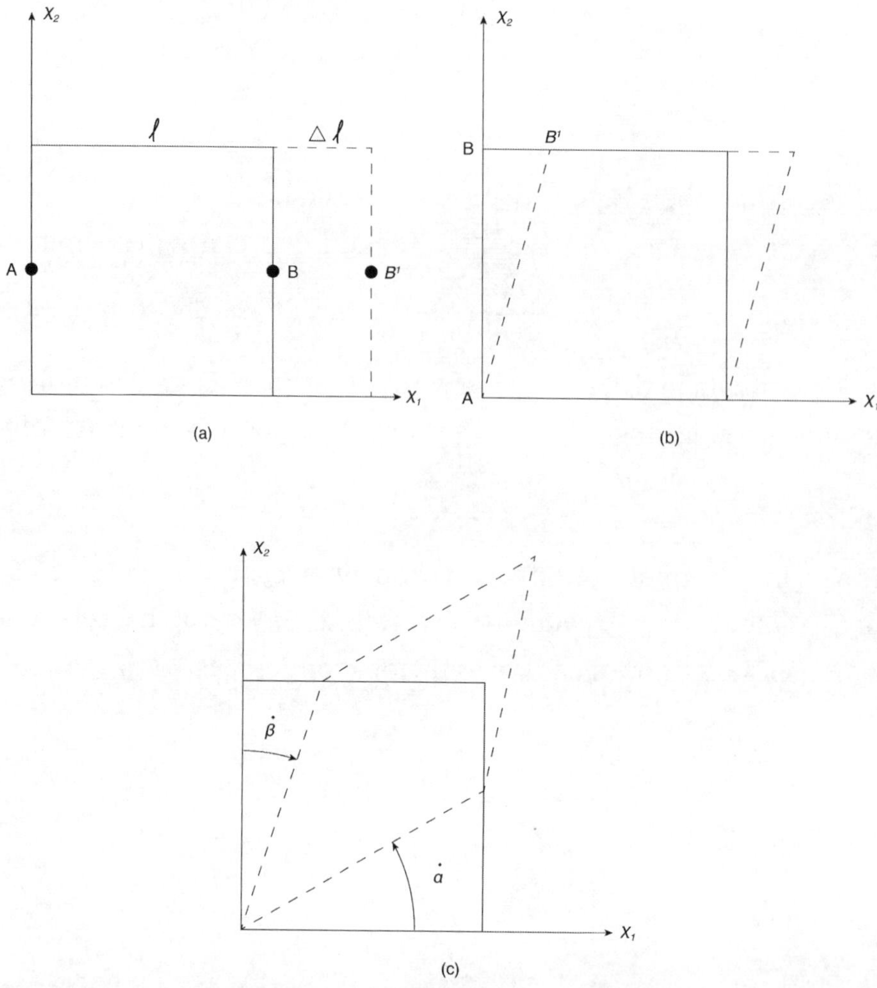

Fig. 7.2 Strain rate deformation showing both (a) linear deformation rates and (b) angular deformation rates; in (c) the net angular deformation rate is zero when = $\dot{\alpha} = \dot{\beta}$.

Now consider Fig. 7.2 (b), which shows the shear strain rate. Or the rate of strain caused by a shearing force. In this case two points on the element surface, A and B, will have a change of orientation relative to a fixed coordinate system. So points A-B will move to A-B' if the x_1 component of velocity increases in the x_2 direction. As shown this can be interpreted as a rotation of the line connecting A-B. If distance B-B' is length Δl and the line length A-B = $dx_2 = l$ then:

$$\frac{\Delta l/l}{\Delta t} = \frac{\partial u_1}{\partial x_2} \tag{7.4}$$

We can interpret this as a rotation, or angular strain rate, which in this case is about the x_3 axis.

Similarly there can be a rotation about x_3 caused by rate, $\frac{\partial u_2}{\partial x_1}$. This can be repeated to obtain angular strain rates about each of the other axes as well.

These results provide the basis for the definition of the strain rate tensor, $\dot{e}_{ij}$:

$$\dot{e}_{ij} = \frac{1}{2}\left(\frac{\partial u_i}{\partial x_j} + \frac{\partial u_j}{\partial x_i}\right) \tag{7.5}$$

Notice that when i=j the strain rate tensor consists of components of the dilation rate, for instance $\frac{\partial u_1}{\partial x_1}$, etc. When $i \neq j$ we see the addition of the two terms causing rotation about any given axis. For instance if i=1 and j=2 we obtain rotation about the x_3 axis as shown above. But also, if i=2 and j=1 we also obtain rotation about the x_3 axis. In fact when added to calculate $\dot{e}_{ij}$ these resultant two cases are identical in value. This implies that the strain rate tensor is symmetric, $\dot{e}_{ij} = \dot{e}_{ji}$. Keep in mind that when the velocity derivative component is negative we have rotation in the opposite direction. The sign convention is that counter-clockwise rotation is positive. The inclusion of ½ is a matter of convenience as we will see later, but also can be thought of as providing an average of the two terms.

The general strain rate tensor, $\dot{e}_{ij}$, consists of all possible combinations of velocity spatial derivatives, and contains 9 elements. The diagonal terms are "linear deformation rates" and the off diagonal terms are the "shearing deformation rates". Notice that there are really only six definitive terms, since this is a symmetric tensor.

Recall that vorticity, is a measure of rotation rate as well. We need to distinguish this from the strain rate tensor. Consider the case of ω_3:

$$\omega_3 = \left(\frac{\partial u_2}{\partial x_1} - \frac{\partial u_1}{\partial x_2}\right) \tag{7.6}$$

First we notice the sign difference for the second term in the parenthesis when compared with the strain rate tensor. What is the consequence of this? If rotation caused by $\frac{\partial u_2}{\partial x_1}$ is counter clockwise because this derivative is positive, and the rotation caused by $\frac{\partial u_1}{\partial x_2}$ is negative when this derivative value is positive, then applying this to the above definition of ω_3 results in a net positive rotation (counter clockwise) about the x_3 axis. Or said another way, the vorticity is the sum of the two contributions of positive rotation rate about a given axis. Notice that vorticity has only three terms needed to define it (hence it is a vector). The diagonal terms are identically zero and the off diagonal terms are just of opposite sign (an antisymmetric tensor).

It is then possible to have zero vorticity (the two terms in the vorticity definition cancel) but still have a positive angular deformation. Consider the case of $\frac{\partial u_2}{\partial x_1} = \frac{\partial u_1}{\partial x_2}$ where the rotation rate is zero but the net angular deformation rate is not, and equal to twice the value of each component. This addition is shown schematically for a fluid element in Fig. 7.2 (c) where the net rotation about the x_3 axis would be zero if $\dot{\alpha} = \dot{\beta}$, but we see deformation of the fluid element.

We do one last manipulation. Suppose we identify the velocity gradient tensor as $\frac{\partial u_i}{\partial x_j}$. This has nine elements of all of the partial derivatives of the three components in the three coordinate directions. This second order tensor can be written as:

$$\frac{\partial u_i}{\partial x_j} = \frac{1}{2}\left(\frac{\partial u_i}{\partial x_j} + \frac{\partial u_j}{\partial x_i}\right) + \frac{1}{2}\left(\frac{\partial u_i}{\partial x_j} - \frac{\partial u_j}{\partial x_i}\right) \tag{7.7}$$

The first term on the right hand side is the deformation rate tensor and the second term is ½ of the vorticity, which (when including the ½) is identified as the rotation rate tensor. The latter can be thought of as the average rotation about any given axis. So the velocity gradient tensor consists of two parts, that due to the strain rate and that due to the vorticity (or rotation rate).

At this point we will construct a fairly simple model for the relationship between viscous stress and deformation rate to be as follows:

$$\tau_{ij} \sim \dot{e}_{ij}$$

This implies that the rate of deformation that occurs in a fluid element is directly proportional to the stress applied to that element. If we make the assumption that there is a proportionality factor equal to 2 μ, then we can write this as:

$$\tau_{ij} = \mu\left(\frac{\partial u_i}{\partial x_j} + \frac{\partial u_j}{\partial x_i}\right) \tag{7.8}$$

It is obvious why the factor 2 was introduced. Notice that the relationship does not include any of the rotational components of the velocity gradient tensor, only the deformation rate components. The physical interpretation is that viscous stress is related to the degree of deformation occurring in a fluid element, not its rotation rate.

Looking at the normal stress components, when $i = j$, we see that these terms result in linear deformation. As noted above these are elongation or contraction rates of a fluid element as it flows over time. Adding all three dimensions we get the net volume change of the fluid element over time. Consequently, according to the model of Eqn. (7.8) the viscous forces resulting in normal viscous stress components are responsible for these linear deformation rates. These stresses are the normal, or diagonal, components of τ_{ij}.

The shearing contributions of τ_{ij} are the off diagonal terms and are related through Eqn. (7.8) to the off diagonal terms of the strain rate tensor, $\dot{e}_{ij}$. These terms result in an angular deformation rate that distorts the original shape of the fluid element over time. Again, according to eqn. (7.8) this angular deformation rate, due to shearing viscous stresses, has a magnitude determined by the parameter, μ.

The parameter μ is a fluid property and can take on many forms depending on the characteristics of the fluid itself. This property is the fluid dynamic viscosity and may be a function of the state of the fluid (say its pressure or temperature), but it may also be a function of the stresses applied to the fluid. A Newtonian fluid is one that can have state varying viscosity (say like an oil that has decreasing viscosity with increasing temperature) but has a constant value of viscosity with changing stress within the fluid. These fluids have a linear relationship between viscous stress and deformation rate.

A dilatant fluid is described as shear thickening. That is to say as a shearing stress is applied to the fluid the viscosity increases. This implies that increasing stresses will result in decreasing deformation rates, consequently not having a linear relationship. Mixtures of cornstarch and water can behave as a dilatant fluid. In contrast, a fluid with a decreasing viscosity with increasing applied stress is called pseudo-plastic, or shear thinning, fluids. Examples of this type of behavior is ketchup, some paints, ink, and others. Fig. 7.3 illustrates the stress-strain rate relationship for various types of fluids where the slope is a measure of viscosity. For the remainder of this book we will deal with Newtonian fluids. Keep in mind, however, functional relationships can be used for non-Newtonian fluid viscosities, whereas we will use constant values in the problems we examine.

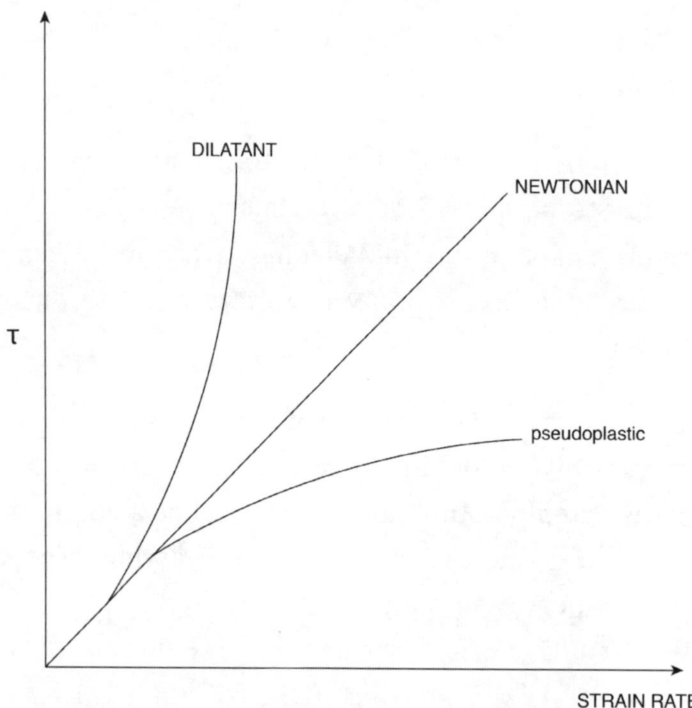

Fig. 7.3 Illustration of the relationship between stress and strain rate for different fluids; shear thickening fluids have increasing viscosity (slope) with increasing stress (dilatant fluids) whereas shear thinning fluids have decreasing viscosity with increasing stress (pseudoplastics); Newtonian fluids have a constant viscosity with strain rate, or a linear relationship.

Navier-Stokes Equation

The Navier-Stokes equation is an extension of Euler's equation with the addition of the viscous forces. The Euler equation is written as:

$$\rho\frac{Du_i}{Dt} = \frac{\partial P}{\partial x_I} + \rho g_i$$

which is a balance of mass times acceleration per volume on the left hand side with the net force due to pressure gradients and the gravitational body force, both per volume of fluid. It should be noted that if we combine the pressure terms with the viscous stress terms we obtain the combined total stress expression:

$$\sigma_{ij} = -P\delta_{ij} + \tau_{ij} \tag{7.9}$$

We use the Kronecker delta, δ_{ij}, which is equal to one when i=j and otherwise equal to zero.

This conveniently indicates that P is a normal compressive stress component. The negative sign indicates that pressure is compressive.

An additional force is introduced to account for forces that result in volume changes of a given fluid element. This is modeled as follows. As noted above the rate change of fluid volume per original volume is given by $\frac{\partial u_k}{\partial x_k}$, where there is summation over index k. Assuming that the force required to change the volume is a compressive normal force, and proportional to the volume change, we can write this force as:

$$\lambda \frac{\partial u_k}{\partial x_k} \delta_{ij}$$

where λ is a proportionality constant dependent on the fluid. This is added to the above stress expression to yield:

$$\sigma_{ij} = -P\delta_{ij} + \tau_{ij} + \lambda \frac{\partial u_k}{\partial x_k} \delta_{ij} \tag{7.10}$$

The net force per volume of fluid due to the stress can be written as:

$$\frac{\partial \sigma_{ij}}{\partial x_j} \tag{7.11}$$

Recall that there is a summation over the index j indicating that this expression is in a vector in the i direction. Inserting Eqn. (7.8) for the stress term in Eqn. (7.10), and then insert this result into Eqn. (7.11) results in the net stress force on a fluid element. This is then added to the Euler equation keeping in mind that we have already included the pressure term with the stresses. The result is:

$$\rho \frac{Du_i}{Dt} = \frac{\partial \sigma_{ij}}{\partial x_j} + \rho g_i = -\frac{\partial P}{\partial x_j} \delta_{ij} + \frac{\partial}{\partial x_j} \left(\lambda \frac{\partial u_k}{\partial x_k} \right) \delta_{ij} + \rho g_i + \frac{\partial}{\partial x_j} \left(\mu \left(\frac{\partial u_i}{\partial x_j} + \frac{\partial u_j}{\partial x_i} \right) \right) \tag{7.12}$$

This equation has three velocity components and the pressure as unknowns. Two fluid parameters, μ and λ are also included. The parameter λ is known as the coefficient of bulk viscosity, and is representative of the force required to change the fluid volume.

For an incompressible fluid, since $\frac{\partial u_k}{\partial x_k} = 0$, (with summation) this term disappears.

For a compressible fluid Stokes hypothesis is that:

$$\lambda = -\frac{2}{3}\mu \tag{7.13}$$

This comes with some theoretical justification and has been shown over a wide range of conditions to be reasonably valid, but shown not to be valid for extreme pressures. Since in this book we restrict ourselves to incompressible flows, this is not an issue and we will neglect this term. Also, for incompressible flows with constant viscosity we can simplify the last (viscous) term in Eqn. (7.12) as:

$$\frac{\partial}{\partial x_j}\left(\mu\left(\frac{\partial u_i}{\partial x_j}+\frac{\partial u_j}{\partial x_i}\right)\right) = \mu\frac{\partial^2 u_i}{\partial {x_j}^2} + \mu\left(\frac{\partial}{\partial x_i}\left(\mu\frac{\partial u_j}{\partial x_j}\right)\right) = \mu\frac{\partial^2 u_i}{\partial {x_j}^2} \tag{7.14}$$

wherein the second term in the middle expression we have reversed the order of differentiation of x_i and x_j and then invoked the incompressible condition leaving only the first term, which is the Laplace of the velocity vector.

The final form of the Navier-Stokes equation that we will use for incompressible constant viscosity (Newtonian fluid) flows is:

$$\rho\frac{Du_i}{Dt} = \rho\frac{\partial u_i}{\partial t} + \rho u_j\frac{\partial u_i}{\partial x_j} = -\frac{\partial P}{\partial x_j}\delta_{ij} + \rho g_i + \mu\frac{\partial^2 u_i}{\partial {x_j}^2} \tag{7.15}$$

The left hand side has been expanded to show the local and convective acceleration terms and on the right hand side we have force contributions from pressure gradients, gravity and viscous forces (normal and shearing). This equation is coupled with the requirements for incompressible flow:

$$\frac{\partial u_k}{\partial x_k} = 0 \tag{7.16}$$

Notice, since we have incompressible flow it is possible to rewrite the convective acceleration term:

$$\rho u_j\frac{\partial u_i}{\partial x_j} = \rho\frac{\partial\left(u_i u_j\right)}{\partial x_j} \tag{7.17}$$

Sometimes this may be referred to as the "conservative form" of the Navier-Stokes equation since it takes advantage of the conservation of mass for incompressible flow.

We can also rewrite the Navier-Stokes equation using the definition of vorticity. First the convective acceleration term invokes the vector identity mentioned, in an earlier chapter in discussion of the generalized Bernoulli equation, that is written as:

$$\rho u_j \frac{\partial u_i}{\partial x_j} = \rho \left(\frac{\partial \left(\frac{1}{2} u_j u_j \right)}{\partial x_i} + \varepsilon_{ijk} \omega_j u_k \right) \tag{7.18}$$

The viscous terms can also use the following vector identity:

$$\mu \frac{\partial^2 u_i}{\partial {x_j}^2} = -\mu \varepsilon_{ijk} \frac{\partial \omega_k}{\partial x_j} \tag{7.19}$$

This last term is the viscosity times the curl of the vorticity. Notice what happens when the flow is irrotational. The second term of Eqn. (7.18) is zero and the viscous stress term, Eqn. (7.19), is also zero. It is left for the reader to see that under these conditions one can arrive at the simplified Bernoulli equation when integrated between any two points within the flow field. If the flow is inviscid the last term of Eqn. (7.15) is zero, but the flow is not necessarily irrotational since the vorticity also enters into the equation in the convective acceleration term. So, an irrotational flow does imply an inviscid flow but an inviscid flow does not imply irrotational flow. Lastly, the viscous term vanishes if the curl of the vorticity is zero, it is not necessary that the flow be irrotational.

Examples of Exact Solutions to Navier-Stokes Equation

1. *Steady Flow Between Parallel Plates or in a Circular Pipe*

For fully developed, steady flow between parallel plates shown in Fig. 7.4 using Cartesian coordinates with flow in the x_1 direction the reader should verify that there is no acceleration and the Navier-Stokes equation reduces to a balance of forces. If the flow is horizontal there is only pressure and viscous forces that balance. If the flow is vertical with no pressure driving the flow then there is a balance of gravitational forces and viscous forces. For plates with a separation distance of 2h with the x_1 axis along the centerline between the plates, as shown in the figure, the form of the velocity distribution in both cases is identical. Note that since there is only the u_1 velocity component then $\frac{\partial u_1}{\partial x_1} = 0$ and we say the flow is fully developed, with no changes in the flow direction and no convective acceleration. For steady flow there is no local acceleration.

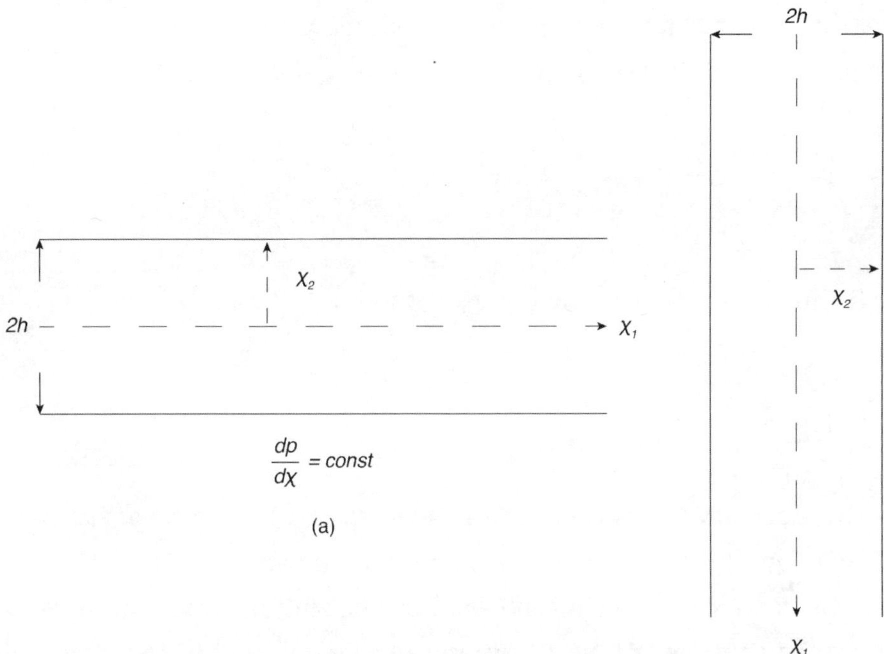

Fig. 7.4 Illustration of steady flow between parallel plates, left is a constant pressure gradient driven flow and on the right is a gravity driven flow; the x_1 coordinate is in the flow direction and x_2 is normal to the flow.

Consider horizontal flow, so no body force, to determine the velocity distribution we first examine the pressure gradient term for this case. Since there is no acceleration the pressure term is balanced by the viscous terms. The resulting Navier–Stokes equation is:

$$0 = -\frac{\partial P}{\partial x_j}\delta_{ij} + \mu\frac{\partial^2 u_i}{\partial {x_j}^2} = \frac{dP}{dx_1} + \mu\frac{d^2 u_1}{dx_2^2}$$

To explain why we can write the terms as ordinary derivatives consider the following. Since the acceleration is zero we can say that $u_1 \neq f(x_1)$ which is also found from the conservation of mass noting that $u_2 = u_3 = 0$. Based on this we can say that the viscous term is not a function of x_1, so therefor the pressure term is not a function of x_1 either. At most u_1 is a function of x_2 . One can show that dP/dx_1 is not a function of x_2 from the x_2 momentum equation. So if the change of P along the flow is not a function of x_2 or x_1, then dP/dx_1 is a constant.

By applying the no-slip boundary condition at the wall surfaces, x_2 = +/- h the result is:

$$u_1 = C\left(1 - \frac{{x_2}^2}{h^2}\right) \tag{7.20}$$

where $C = -\frac{h^2}{2\mu}\frac{dP}{dx_1}$ for pressure driven horizontal flow and $C = \frac{\rho g h^2}{2\mu}$ for gravity driven vertical downward flow.

Given this velocity field the vorticity distribution can be determined and since the velocity is parabolic the vorticity distribution will be piecewise linear across the flow with zero value at the center, and maximum value at the surface, $\omega_3 = -\frac{2C}{h^2}x_2$.

The same flow can be considered for flow in a round constant area horizontal pipe. This is often called Hagen-Poiseuille flow after two experimentalists who developed this relationship in the 1800s for steady pipe flow in the form of flow rate versus pressure drop. In this case the Navier-Stokes equation is written in cylindrical coordinates as shown in Table 7.1 and then reduced for steady, constant area, incompressible, constant property flow. Again, the flow has no acceleration for steady, constant area flow and the result is a balance of pressure and viscous forces in the z, or axial direction. The pressure gradient along z is a constant similar to the result for flow between parallel plates. The resulting simplified equation is integrated across the flow (r coordinate) using the no-slip boundary condition at $r = R$ (R is the pipe radius) and also noting that the velocity is a maximum at the centerline where the velocity derivative in r is zero. The result is:

$$v_z = -\frac{R^2}{4\mu}\frac{dP}{dz}\left(1 - \frac{r^2}{R^2}\right) \tag{7.21}$$

For all of these internal flows the flow rate can be obtained by integrating the velocity distribution across the flow area. For instance, for a volumetric flow rate of $\dot{Q}$ and writing the pressure gradient $(-dP/dx_1)$ as $\Delta P/L$ since it is constant (notice the sign change since ΔP is the upstream pressure minus the downstream pressure) and L is the total length along x_1. We have for steady horizontal pressure driven flow between parallel plates:

$$\dot{Q} = \frac{2h^3\Delta P}{3\mu L}1\,(\text{per distance } x_3) \tag{7.22}$$

For steady flow in a pipe:

$$\dot{Q} = \frac{\pi R^4 \Delta P}{8\mu L} \tag{7.23}$$

It is straight forward to evaluate similar relationships for gravity driven flows by replacing the pressure gradient with ρg.

There are a number of other steady flows that can also be solved for analytically from the Navier-

Stokes equation. For instance, steady axial flow in an annulus, either pressure or gravity driven; flow in the gap between two rotating cylinders that have different rotation rates; flow between parallel plates where one plate is moving relative to the other with or without a pressure gradient or gravity driving the flow, and a few others. All of these flows have no local or convective acceleration terms and the basic governing equation is a balance of forces. It is possible to solve some flows with acceleration as shown next.

2. *Suddenly Accelerated Plate (Stokes First Problem, also known as the Rayleigh Problem)*

When a flat plate is immersed in a large body of fluid, and the plate is suddenly accelerated to some fixed, constant velocity the fluid surrounding the plate will be set into motion. This general problem is known as Stokes first problems, named for Sir George Stokes the well-known mathematician, for whom the Navier-Stokes equation is also named. There are variations on this problem, such as the flow field set up in a fluid when the plate oscillates back and forth sinusoidally, Stokes second problem. Here we will look at the solution one can obtain for the first problem. It will introduce the idea of a similarity formulation of the problem and provide insight on how to nondimensionalize results. It is also useful in the interpretation of boundary layer flows that we will examine in a later chapter.

This is an idealized problem where the plate is assumed to accelerate instantaneously to some velocity $U = U_p$ and remains at that velocity. Obviously there is some finite time for this to happen, but we ignore that (short) time effect. Fig. 7.5 illustrates the coordinate system and the flow. By symmetry we only need to include one half of the flow field, on top, and we assume that the plate is infinite in extent in the x_1, x_3 directions. The flow is only in the x_1 direction, to the right. There are no forcing functions is the x_2 and x_3 directions so we take u_2 and u_3 to be zero. Finally the fluid is assumed to extend infinitely far in the +/- x_2 directions.

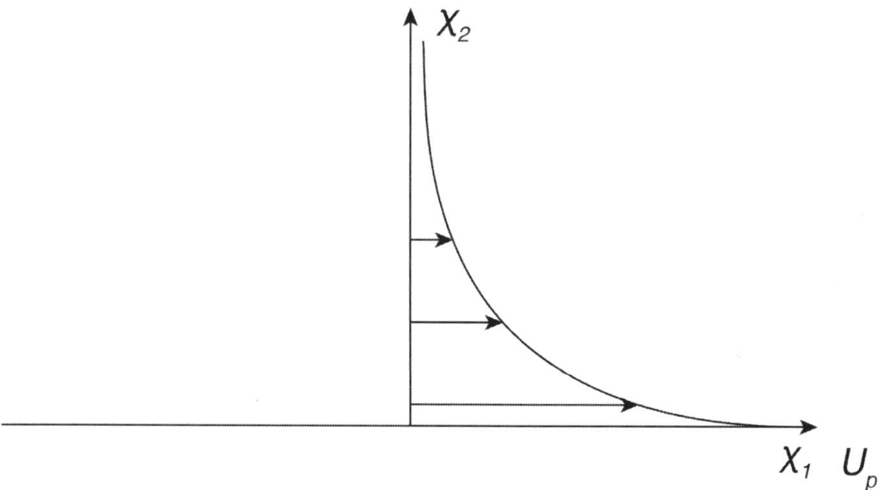

Fig. 7.5 Velocity profile associated with a horizontal plate with near instantaneous acceleration to a constant velocity of U_p at time zero; fluid above the plate is set into motion by viscous forces; the extend of the flow domain above the plate increases with time.

This problem may model the start up of a plate with lubricating oil between the plate and some surrounding surfaces, at least for early times before the flow is influenced by the surrounding surfaces (we will be able to determine this time from our solution).

The analysis begins using the x_1 direction Navier-Stokes equation and eliminating all zero valued terms. Assuming the flow is governing by viscous forces between the surface and fluid, and neglecting any establish pressure variation within the flow we have the simplified form:

$$\frac{\partial u_1}{\partial t} = \nu \frac{\partial^2 u_1}{\partial x^2} \tag{7.24}$$

where $\nu = \mu/\rho$ which is called the kinematic viscosity and has SI units of (m^2/s). This is obtained by dividing through by the fluid density. The associated boundary and initial conditions are:

No-slip: $u_1 = U_p \; at \; x_2 = 0$

No flow far from the surface: $u_1 = 0 \; at \; x_2 \to \infty$

Initially no flow: $u_1 = 0 \; at \; t = 0 \; for \; all \; x_2$

This is what is called a "diffusion" problem where velocity (or momentum per mass) of the fluid diffuses away from a source of momentum (the wall friction force). Fick's law of diffusion states that the diffusion rate, or flux of a quantity, is proportional to the local gradient of the quantity. As we will see this diffusion process is directly linked to the fluid viscosity.

The above governing equation can be scaled in several different ways. First let's use dimensional analysis with the condition that the velocity at any point in space and time within the fluid near the plate is a function of the following list of variables

$$u = (U_p, x_2,\ t,\ \nu) \tag{7.25}$$

where we have used u instead of u_1 since there is only one velocity component to consider. This list of variables should be obvious from the governing equation, (7.24) and the associated initial and boundary conditions. Now using dimensional analysis we arrive at the following three nondimensional Pi groups:

$$\Pi_1 = \frac{u}{U_p}, \Pi_2 = \frac{x_2}{\sqrt{\nu t}}, \Pi_3 = \frac{x_2}{U_p t} \tag{7.26}$$

However, Π_3 provides no new information since all of the variables appear in the other two Pi groups. Hence we can conclude that

$$\Pi_1 = f\,(\Pi_2) \text{ or } \frac{u}{U_p} = f\frac{x_2}{\sqrt{\nu t}} \tag{7.27}$$

This implies that the velocity scaling for this problem is U_p and the length scale is $\sqrt{\nu t}$. Notice here that the length scale is a function of time. That is to say that the extent of the length scale increases over time, and is not a constant. We can think of this length scale as roughly the distance over which diffusion is expected to occur, which is time dependent. For convenience in the mathematics we will add an arbitrary "2" to the length scale and express it as:

$$\frac{u}{U_p} = f(\frac{x_2}{2\sqrt{\nu t}}) \tag{7.28}$$

We have two variables, $f = \dfrac{u}{U_p}$ and $\eta = \dfrac{x_2}{2\sqrt{\nu t}}$. And we seek a solution of $f(\eta)$. The next step is to substitute our nondimensional variables into the governing equation and initial and boundary conditions. This needs to be done carefully since one of our scales (length) is a function of time. The reader is encouraged to go through the following transformation using the chain rule:

$$\frac{\partial u}{\partial t} = \frac{\partial (fU_p)}{\partial \eta} \cdot \frac{\partial \eta}{\partial t} = U_p \frac{\partial f}{\partial \eta}(-\frac{\eta}{2t})$$

$$\frac{\partial u}{\partial x_2} = U_p \frac{\partial f}{\partial \eta} \cdot \frac{\partial \eta}{\partial x_2} = U_p \frac{\partial f}{\partial \eta} \frac{1}{2\sqrt{\nu t}}$$

$$\frac{\partial^2 u}{\partial {x_2}^2} = \frac{\partial}{\partial x_2}\left(U_p \frac{\partial f}{\partial \eta} \cdot \frac{\partial \eta}{\partial x_2}\right) = U_p \frac{\partial f}{\partial \eta} \frac{\partial^2 \eta}{\partial {x_2}^2} + U_p \frac{\partial \eta}{\partial x_2} \frac{\partial^2 f}{\partial \eta^2} \frac{\partial \eta}{\partial x_2}$$

$$\frac{\partial^2 u}{\partial {x_2}^2} = U_p \left(\frac{1}{2\sqrt{\nu t}}\right)^2 \frac{\partial^2 f}{\partial \eta^2}$$

Inserting the first and the last of this set of equations in Eqn. (7.24) the result is:

$$\frac{\partial^2 f}{\partial \eta^2} + 2\eta \frac{\partial f}{\partial \eta} = 0 = f'' + 2\eta f' \tag{7.29}$$

where the last expression is used to write derivatives as primes (f'' is the second derivative and f' is the first derivative.) Notice here that the derivatives can actually be ordinary derivatives since as shown in (7.27) f is only a function of η. Importantly here the second order partial differential equation in two variables is transformed into a second order ordinary differential equation, which in general has a much easier solution.

The next step is to transform the boundary conditions as follows:

$$\begin{aligned}
&\text{at } x_2 = 0 \;\; \eta = 0 \text{ and } u = U_p \text{ so } f(0) = 1 \\
&\text{as } x_2 \to \infty \;\; \eta \to \infty \text{ and } u \to 0 \text{ so } f(0) = 0 \\
&\text{at } t = 0 \;\; \eta \to \infty \text{ and } u \to 0 \text{ so } f(0) = 0
\end{aligned} \tag{7.30}$$

The last two conditions are identical, so we end up with two boundary conditions for η which is what we expect since we have a second order equation. We are left with the governing equation (7.29) and the boundary conditions (7.30). It turns out that there is an analytical solution. Letting:

$$g = f'$$

then the governing equation is:

$$\frac{g'}{g} = -2\eta$$

this can be integrated to:

$$g = f' = C_1 e^{-\eta^2}$$

which we integrate again to:

$$f = C_1 \int_0^{\eta} e^{-\eta^2}\, d\eta + C_2 \tag{7.31}$$

The first term on the right hand side is a straightforward integral evaluation up to a variable value η. This can be replaced by the Gaussian Error Function, $erf(\eta)$ which is shown in Fig (7.6) as:

$$erf\,(\eta) \;=\; \frac{2}{\sqrt{\pi}} \int_0^{\eta} e^{-\eta^2}\, d\eta$$

so that:

$$f = \frac{\sqrt{\pi}}{2} C_1 erf\,(\eta) \;+ C_2$$

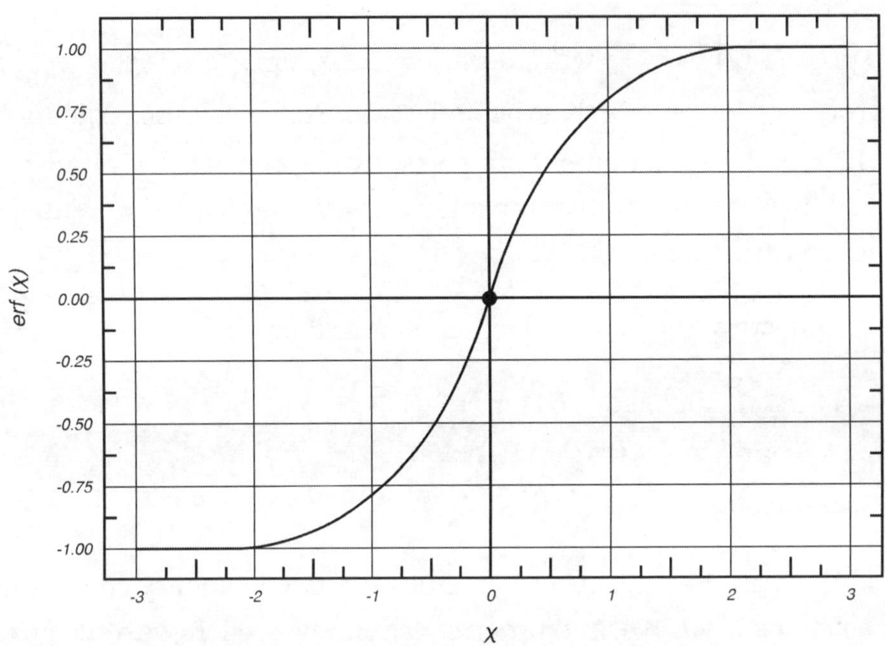

Fig. 7.6 Gaussian Error Function, erf(x), where x is the argument; the function is normalized to vary between -1 and +1 and passes through zero at $x = 0$.

We are left to evaluate the constants from the two boundary conditions for f.

Since $erf(0) = 0$ and $erf(\infty) = 1$ then

$C_1 = -\dfrac{2}{\sqrt{\pi}}$ and $C_2 = 1$ and the final form of the equation is:

$$f = \frac{u}{U} = 1 - erf(\eta) \text{ and } f' = -\frac{2}{\sqrt{\pi}} e^{-\eta^2} \tag{7.32}$$

The results of Eqn. (7.32) are plotted in Fig. 7.7. This solution is a "similarity solution" in that the results collapse, in this case, into a single variable, η, which depends on the two independent variables, x_2 and t. There are two parameters affecting the result of the local velocity, the boundary condition, U_p, and the fluid property, ν. To evaluate a given set of conditions one does the following. (1) Identify the physical space and boundary conditions of interest, (2) transform variables such as values of space or time into associated values of the variable η. (3) Use Eqn. (7.32) to determine the value of f versus η and the desired η value. (4) Transform f into the desired velocity u.

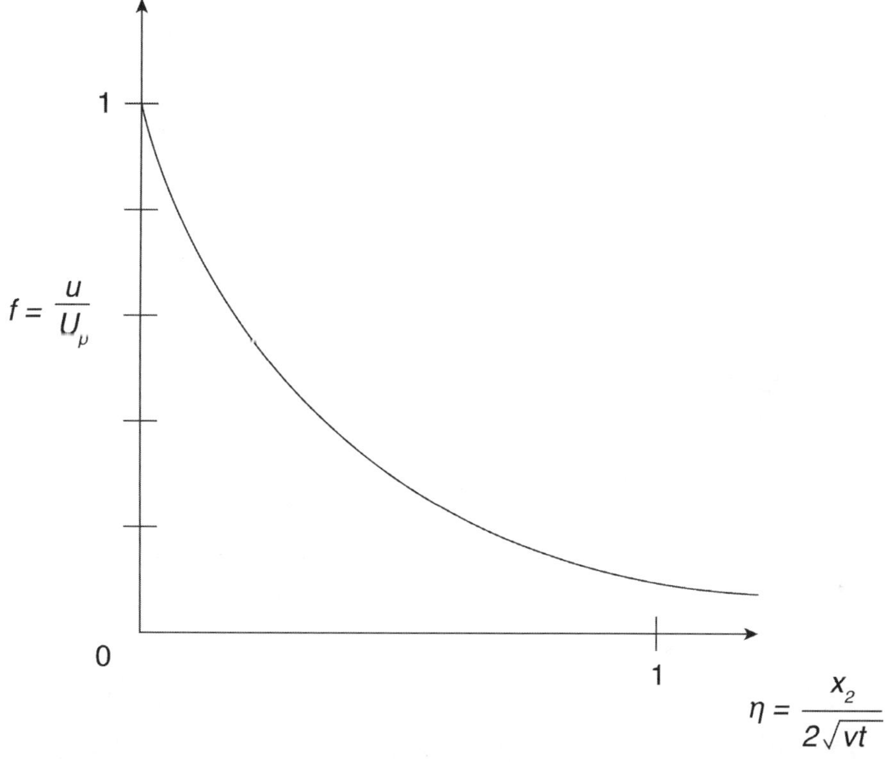

Fig. 7.7 Nondimensional result of f' versus η.

Besides determining the local velocity in the fluid these results are also useful in determining the transient wall shear stress, or force per unit area of the plate. Given in Eqn. (7.32) is the derivative,

f'. This provides information of the velocity derivative, $\dfrac{\partial u}{\partial x_2}$. Since there is only one velocity component and it is only a function of x_2 we can write the shear stress as:

$$\tau = \mu\left(\frac{\partial u_2}{\partial x_1} + \frac{\partial u_1}{\partial x_2}\right) = \mu\frac{\partial u_1}{\partial x_2} = \mu\left(\frac{Uf'}{2\sqrt{\nu t}}\right) \tag{7.33}$$

where f' is the derivative $\dfrac{\partial f}{\partial \eta}$. The force on the surface of the plate with an area of LS, (where L is the distance along the plate in the direction of motion and S is the width, or span of the plate) due to its motion is:

$$F_p = \int_0^L \tau_s dx_1 S \tag{7.34}$$

where the shear stress, τ_s, is evaluated at the surface where x_2 = 0. Evaluating (7.34) using (7.33) at $\eta = 0$, with $f'(0) = -\dfrac{2}{\sqrt{\pi}}$ results in:

$$\tau_s = -\frac{\mu U_p}{\sqrt{\pi \nu t}}$$

and the force

$$F_p = -\frac{\mu U_p LS}{\sqrt{\pi \nu t}} \tag{7.35}$$

Notice that the wall stress decreases with time as momentum diffuses away from the surface. The force therefore also decreases with time. At time t = 0 the equations have a problem, since we said the acceleration was initially infinite.

We can define the diffusion distance way from the surface as the distance x_2 to where the velocity decays to 1% (small arbitrary value) of the plate velocity. So setting f = 0.01 we see that η = 1.85 and setting $x_2 = \delta$ at this point we have:

$$\delta = 3.7\sqrt{\nu t} \tag{7.36}$$

Notice that the diffusion distance increases with time as well as kinematic viscosity. The greater the viscosity the further is the distance, δ, at any given time. It is independent of the plate velocity.

One last item on this seemingly simple problem that will help in the interpretation of boundary layer flows. This flow has vorticity, but only aligned with the x_3 direction, as can be seen by calculating the local vorticity distribution:

$$\omega_3 = \left(-\frac{\partial u}{\partial x_2}\right) = -\frac{Uf'}{2\sqrt{\nu t}} \tag{7.37}$$

The circulation is the area integral of the vorticity, so extending an integration from the surface at $x_2 = 0$ to δ, for a distance L along x_1, then the total circulation per distance x_3 becomes:

$$\Gamma = \int_0^\delta -\frac{U_p f'}{2\sqrt{\nu t}} dx_2 L = \int_0^\infty -U_p f' d\eta L = -U_p L$$

where for x_2 greater than δ the value of $f' = 0$ so there is no contribution to Γ.

Interestingly, the total circulation remains constant overtime, but the area grows since δ increases with time. The diffusion process merely spreads the vorticity over an increasing area, for increasing time and all of the vorticity is generated with the initial acceleration of the surface.

VIII. Boundary Layer Flows

In this chapter, we discuss the physical attributes associated with boundary layer flows. The governing equations are developed from the Navier-Stokes equation. The laminar boundary layer flow characteristics and interpretation of the associated forces generated by the flow are presented and discussed. This flow is a viscous dominated flow, rather than a pressure dominated flow. We will not explore the details associated with boundary layers on curved surfaces or surfaces that have flows influenced by an existing pressure gradient. Some physical insight into these conditions will be presented, however, and the concept of flow separation is introduced.

The pioneering work of Ludwig Prandtl soon after the beginning of the 20th century lays most of the groundwork for boundary layer analysis. His work has lead to an entire field of study that allows a combination of viscous dominated flows with inviscid flows. Prandtl used a very detailed combination of experimental and analytical methods to approach a very complex problem – the influence of fluid friction on streamline flows. He was very interested in wing theory and how to model real wing flows and their associated forces of lift and drag. The concept of a thin airfoil theory was developed and applied to three-dimensional wings. Here we will look at the basic physics associated with boundary layer theory and how these flows can be modeled for the laminar flow conditions. Turbulent flows are introduced in a later chapter.

Physical Attributes of a Boundary Layer

A boundary layer flow is defined to be the region of a larger flow field that is next to the surface and has significant effects of wall frictional forces. Since the region of interest is near the surface and the surface is assumed to be impervious to the flow then the velocity is nearly parallel to the surface. The general flow is shown in Fig. 8.1. Flow is from left to right and there is an upstream "leading edge" where the surface begins. At the leading edge (defined as the coordinate system origin), the flow immediately next to the surface begins to experience frictional forces due to the no slip boundary condition.

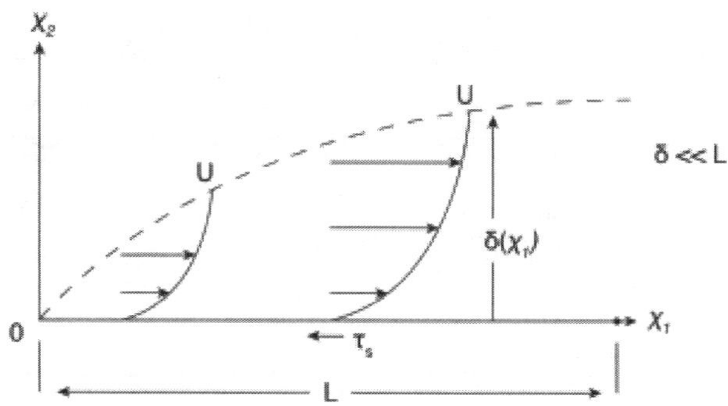

Fig 8.1 Illustration of boundary layer flower characteristics and coordinate system.

The fluid outside of this thin layer is directly unaffected by the wall friction. As the flow goes downstream the slower moving fluid near the surface exerts frictional forces on the fluid further away from the surface. This process is one of diffusion of momentum loss in a direction normal to the surface that results in a reduction of the local fluid velocity. The rate of diffusion depends on the viscosity of the fluid. As this happens the region slowed by friction grows such that the thickness of the boundary layer, δ, increases along the flow direction. At the outer edge of this viscous region, where the flow is no longer affected or slowed by the surface generated friction, the velocity is called the "freestream" value. The boundary layer flow shown in Fig. 8.1 represents a laminar flow condition where the boundary layer thickness extends further away from the surface at locations further along the flow direction. Turbulent flow has a similar trend but has some distinct differences in the velocity distribution and is discussed in a later chapter. Both laminar and turbulent flows must follow the no-slip boundary condition at the surface, resulting in zero relative velocity at the surface. The boundary layer thickness is affected by the turbulence and results in a thicker boundary layer in general.

Although we said that the flow is principally along the surface in the narrow boundary layer, there needs to also be a "wall-normal" velocity component, or u_2 using our coordinate system. The requirement of a non-zero velocity component away from the surface can be shown by using a control volume starting from the leading edge and extending downstream an arbitrary distance x_1 as shown in Fig. 8.2. Note that the control volume has the same area for flow inlet and outlet in the stream-wise direction, that is h is a constant. By applying conservation of mass to this control volume we obtain the following for steady flow:

$$0 = \dot{m}_3 + \dot{m}_2 - \dot{m}_1$$

where the subscripts refer to the different mass flows shown in Fig. 8.2, subscript 1 is the flow coming into the control volume from the left at the leading edge, subscript 2 is the flow going out the right side at distance x downstream and subscript 3 is flow that goes out the top. The flow coming in at the leading edge has a uniform velocity of U over the vertical distance h. The outflow at distance x_1 has been slowed by friction such that the velocity varies from zero at the surface to U at the outer edge of the boundary layer. Integrating this downstream velocity profile over the same vertical distance h will result in a lower mass flow rate than at the leading edge since $\dot{m}_2 < \dot{m}_1$. Consequently, $\dot{m}_3 > 0$ or, there must be a positive outflow, $\dot{m}_3$, resulting in a velocity component out the top of the control volume. The no-slip boundary condition requires that all velocity components are zero at the surface. This implies that the vertical velocity component must start at zero and increase with increasing x_2. Also, streamlines associated with the flow must deflect upwards along the flow direction to account for the nonzero x_2 velocity component, as is shown in Fig 8.2.

The characteristics of the developing velocity profile along the flow direction, x_1, can be used to understand the distribution of the surface shear stress τ_s. Since the boundary layer grows larger in the flow direction, as more and more of the fluid above the surface is slowed by friction, then at any given height, x_2, the velocity in the x_1 direction must decrease going in the x_1 direction. That is $\dfrac{\partial u_1}{\partial x_1} < 0$.

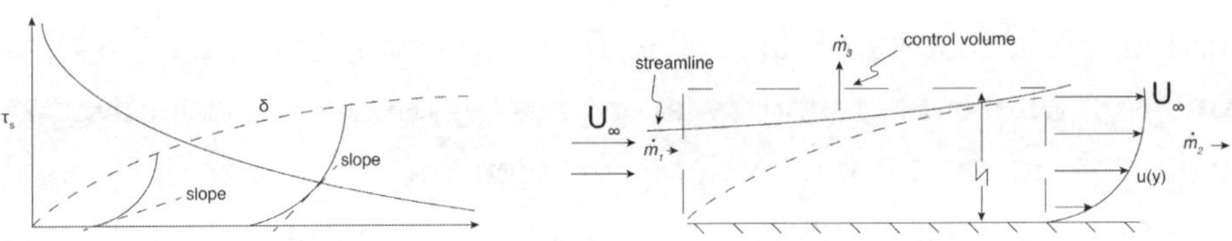

Fig. 8.2 Distribution of the surface shear stress along the surface the shear stress continues to decreases as the boundary layer increases in the flow direction; control volume used to illustrate conservation of mass requirements for wall-normal flow.

The result of this is that the velocity distribution has a decreasing value of $\dfrac{\partial u_1}{\partial x_2}$ near the surface. That is to say, the velocity is always zero at the surface and always extends to U at the edge of the boundary layer, $x_2 = \delta$, so as δ increases the velocity derivatives $\dfrac{\partial u_1}{\partial x_2}$ everywhere within the flow must decrease. Based on this, the value of $\dfrac{\partial u_1}{\partial x_2}$ at the surface decreases in the x_1 direction. Using the definition of the shear stress and noting that $\dfrac{\partial u_1}{\partial x_1} = 0$ at the surface, we see that

$$\tau_s = \mu \left(\frac{\partial u_2}{\partial x_2} \right)_{x_2=0} \tag{8.1}$$

where the subscript $x_2 = 0$ means that it is evaluated at the surface. Since the surface velocity derivative decreases along x_1 then so does the surface shear stress and the qualitative result is the trend of τ_s shown in Fig. 8.2. Notice that the surface shear stress does not reach a constant value, as in pipe flow, since the flow never becomes fully developed, or $\dfrac{\partial u_1}{\partial x_1} \neq 0$.

If the surface is perfectly flat then the freestream flow outside the boundary layer will be nearly parallel with the surface (in the x_1 direction) at velocity U, except for the small upward velocity component described earlier. For parallel straight flow the change of pressure along the flow, and normal to the flow, is zero as we saw in inviscid flow. If we anticipate that the thickness, δ , of the boundary layer is relatively thin (we will show this to be the case later in this chapter when we develop the boundary layer equations) then the change of pressure due to hydrostatic effects between the outer edge of the boundary layer and the surface is small, even if the fluid is a liquid. However, even if hydrostatic pressure rise is not small across the boundary layer the change in pressure along the flow should in fact approach zero near the surface because the change in pressure along the flow is zero just outside of the boundary layer. This is explained by the following. The hydrostatic pressure change is the same between two vertical points evaluated at different locations along x_1. So the difference of pressure in the x_1 direction, $\dfrac{\partial P}{\partial x_1}$ will be identical for positions near the edge of the boundary layer and near the surface. So if $\dfrac{\partial P}{\partial x_1} = 0$ just outside the boundary layer then it is also zero inside the boundary layer. The net effect of this is that for a flat surface the pressure gradient along the flow is essentially zero. We will show this to be true based on the governing equations later in this chapter. We could extend this further for a curved surface and deduce that the pressure gradient for flow over the curved surface is the same outside and inside the boundary layer if the boundary layer is thin.

So a summary of the main attributes we will rely on for the analysis of boundary layers at this point are: (i) the boundary layer is thin, δ >> L, (ii), the wall normal velocity is small but finite, u_2 << u_1, (iii) for a (nearly) flat surface the pressure gradient is zero, $\partial P/\partial x_1 = 0$, (iv) the outer edge of the boundary layer velocity is the freestream velocity, U, which is determined by inviscid flow conditions.

Boundary Layer Equations

The boundary layer equations are a somewhat simplified form of the Navier-Stokes equations based on the physical attributes of the flow, listed above. The flow here is taken to be steady and two-dimensional, with velocity components in x_1 and x_2, which we designated by (u_1, u_2). Note that it is possible to have a more complex unsteady three dimensional version of boundary layer flow that is beyond the scope here. As part of the solution we will show that u_1 >> u_2. We assume the surface is perfectly flat and parallel to the oncoming fluid flow direction so there is no pressure gradient along the flow direction. Finally, we neglect any gravitational forces and assume we have incompressible flow. Strictly speaking the boundary layer equations do not require the pressure gradient to be zero, and can be treated as a parameter that allowed for some curvature of the surface (not flat).'

With the above conditions we write the x_1 direction Navier-Stokes equation as:

$$\rho\left(u_1\frac{\partial u_1}{\partial x_1}+u_2\frac{\partial u_1}{\partial x_2}\right)=\mu\left(\frac{\partial^2 u_1}{\partial {x_1}^2}+\frac{\partial^2 u_1}{\partial {x_2}^2}\right) \tag{8.2}$$

we combine this with the differential conservation of mass for an incompressible flow:

$$\left(\frac{\partial u_1}{\partial x_1}+\frac{\partial u_2}{\partial x_2}\right)=0 \tag{8.3}$$

At this point we have two-dimensional partial differential equations in two variables. It is worth noting here that the pressure gradient in the y direction, across the flow, is reduced to a very small number under the conditions that the u_2 component of velocity is very small compared with u_1, and that the boundary layer is thin so that there is essentially zero hydrostatic pressure variation across the flow. This can be shown by noting that all terms in the x_2 momentum equation containing u_2 become very small and the body force term is small since changes in x_2 are small, therefor the pressure gradient in x_2 must also be small. This process will become clearer in our discussions of scaling of the x_1 momentum equation below.

Next we scale the variables in the governing equation. This means coming up with appropriate scaling for each variable (dependent and independent) that is on the order of the maximum value of the variable. For instance, the characteristic length scale for flow over a flat plate with length L would be just L. Similarly, the characteristic velocity scale for the x_1 velocity component would be the freestream velocity, U. Below is listed the characteristic scales for the various variables (note that the scales may not be constant values as is noted by the choice of scale for x_2):

$$x_1 - L \tag{8.4}$$

$x_2 - \delta$
$u_1 - U_\infty$
$u_2 - unknown$

In general L can be any length of surface. We select δ to be the boundary layer thickness that coincides with this selected length in the analysis below. That is to say that the largest value of δ occurs at $x_1 = L$. But in general δ varies as we move in the x_1 direction. We do not know how large u_2 may be, but we can determine this by examining the conservation of mass, Eqn. (8.3). To do this we use the above scaling to form nondimensional variables, denoted by the superscript "*" by dividing each variable by its scale value:

$$x_1{}^* = \frac{x_1}{L} \tag{8.5}$$

$$x_2{}^* = \frac{x_2}{\delta}$$
$$u_1{}^* = \frac{u_1}{U}$$
$$u_2{}^* = \frac{u_2}{V_s}$$

where V_s is an unknown scale for variable u_2 and is to be determined. Inserting these non-dimensional variables into Equation (8.3), where $u = u^* U$, etc., we obtain:

$$\frac{U}{L}\left(\frac{\partial u_1{}^*}{\partial x_1{}^*}\right) + \frac{V_s}{\delta}\left(\frac{\partial u_2{}^*}{\partial x_2{}^*}\right) = 0 = \frac{\partial u_1{}^*}{\partial x_1{}^*} + \frac{V_s L}{\delta U}\left(\frac{\partial u_2{}^*}{\partial x_2{}^*}\right) \tag{8.6}$$

where the last set of terms is a consequence of dividing the first set of terms through by $\frac{U}{L}$. The coefficient of the first term become one, the coefficient multiplying the second term becomes $\frac{V_s L}{\delta U}$. At this point we apply an order of magnitude analysis of the terms. Both terms have to balance each other in the equation, being of opposite sign. We use the nomenclature $\mathcal{O}(1)$ to indicate that the value of a term is "order one". By proper scaling of the variables we assume that, $\frac{\partial u_1{}^*}{\partial x_1{}^*}$, is $\mathcal{O}(1)$. That is to say, by proper scaling this derivative using scales that are approximately the maximum value of the variable then the resultant term will not be a very large or a very small number, but rather "order one". We make the same assumption for the derivative $\frac{\partial u_2{}^*}{\partial x_2{}^*}$. Consequently since neither of these derivatives are zero, as we have noted previously for the boundary layer, then for this equation to be valid we must have

$$\frac{V_s L}{\delta U} = \mathcal{O}(1)$$

With this we can assign the value of the u_2 velocity scale to be:

$$V_s = U\frac{\delta}{L} \tag{8.7}$$

By doing so then each term is $\mathcal{O}(1)$. Keep in mind that we are not trying to find exact values but at this point trying to determine approximately how large or small the scaling values should be. If we now take Equation (8.7) and insert this into Equation (8.6) we end up with:

$$\left(\frac{\partial u^*}{\partial x^*}\right) + \left(\frac{\partial v^*}{\partial y^*}\right) = 0 \tag{8.8}$$

for the nondimensional representation of the conservation of mass with the scaling of the variables given by the relationships above.

Using our new found scaling for u_2 we continue this same non-dimensional process for the other terms in the Navier-Stokes Equation (8.2). That is, replacing all of the variables with the nondimensional variables, and then dividing the resulting equation by the coefficient of the first term, $\frac{\rho U^2}{L}$ resulting in:

$$u_1{}^*\frac{\partial u_1{}^*}{\partial x_1{}^*} + u_2{}^*\frac{\partial u_1{}^*}{\partial x_2{}^*} = \frac{\mu}{\rho UL}\left(\frac{\partial^2 u_1{}^*}{\partial x_1{}^{*2}} + \frac{L^2}{\delta^2}\frac{\partial^2 u_1{}^*}{\partial x_2{}^{*2}}\right) \tag{8.9}$$

This nondimensional equation, that incorporates the scaling given above, leads to some profound conclusions. First note that the coefficient on the right hand side of the two terms in parentheses, which are the fluid friction terms, is the inverse of the Reynolds number that we define for boundary layer flows: $Re_L = \frac{\rho U_\infty L}{\mu}$. This Reynolds number is based on the freestream velocity and the total length of the surface. We have included a subscript L on the Reynolds number to inform us of the selected length scale. Next we have the coefficient $\frac{L^2}{\delta^2}$ of the last term in parentheses. Since we have assumed that the boundary layer is thin ($\delta \ll L$) then this coefficient will be large. Since each of the derivatives is assumed to be order one through proper scaling, then the entire second term in parentheses must be much larger than the first term, or:

$$\frac{\partial^2 u_1{}^*}{\partial x_1{}^{*2}}$$

<<

$$\frac{L^2}{\delta^2}\frac{\partial^2 {u_1}^*}{\partial {x_2}^{*2}} \tag{8.10}$$

So the frictional forces in the boundary layer are dominated by the term containing $\frac{\partial^2 {u_1}^*}{\partial {x_2}^{*2}}$. This term is the net angular deformation caused by shearing stress. The first term is the linear deformation caused by the normal viscous stress. This result is consistent with the fact that the thin boundary layer will have very large derivatives of velocity in the x_2 direction when compared with the x_1 direction. So neglecting the normal viscous stress contribution the resulting nondimensional form of the boundary layer equation for steady flow over a flat surface is:

$${u_1}^*\frac{\partial {u_1}^*}{\partial {x_1}^*} + {u_2}^*\frac{\partial {u_1}^*}{\partial {x_2}^*} = \frac{\mu}{\rho U L}\left(\frac{L^2}{\delta^2}\frac{\partial^2 {u_1}^*}{\partial {x_2}^{*2}}\right) \tag{8.11}$$

There is one more conclusion that can be arrived at from this scaling. Noting that the shearing contribution derivative, $\frac{\partial^2 {u_1}^*}{\partial {x_2}^{*2}}$, is $\mathcal{O}(1)$ as are each of the nondimensional acceleration terms on the left hand side, then in order for this equation to be balance $\left(\frac{\mu}{\rho U L}\frac{L^2}{\delta^2}\right)$ should also be $\mathcal{O}\left(1\right)$:

$$\frac{\mu}{\rho U L}\left(\frac{L^2}{\delta^2}\right) \sim \mathcal{O}(1)$$

or

$$Re_L \sim \frac{L^2}{\delta^2} \tag{8.12}$$

This last condition implies that for a thin boundary layer to exist the Reynolds number should be large. So for large Reynolds numbers in boundary layer flows with a surface length of L the boundary layer thickness, δ, is proportional to the square root of Re_L which we can write as:

$$\frac{\delta}{L} = \frac{C}{\sqrt{Re_{x_1}}} \tag{8.13}$$

where C is some unknown constant inserted to make the equation quantitatively correct.

Consequently, for very small values of L the boundary layer conditions do not hold since Re_L is small.

The nature of the governing equation given by Eqn. (8.11) is that it is a parabolic partial differential equation. Since it only requires one boundary condition in x_1, which can be selected at $x_1 = 0$ (the leading edge) where the velocity is given as U, then the solution is independent of any downstream boundary condition. This means it can be solved at any arbitrary position x_1 along the surface. This also implies that the result (8.13) is valid by replacing L with location x_1. . Doing this we have a Reynolds number based on any position x_1:

$$Re_{x_1} = \frac{\rho U x_1}{\mu}$$

and Equation (8.13) can be expressed as:

$$\frac{\delta}{x_1} = \frac{C}{\sqrt{Re_{x_1}}} \tag{8.14}$$

Equation (8.14) predicts that the boundary layer grows proportional to the distance ${x_1}^{1/2}$ (recall that there is a factor x_1 contained in the Re_{x_1}); also this will not be valid for very small values of x_1.

Using Eqn. (8.12) the final form of the boundary layer equation for flow over a flat surface is:

$${u_1}^* \frac{\partial {u_1}^*}{\partial {x_1}^*} + {u_2}^* \frac{\partial {u_1}^*}{\partial {x_2}^*} = \frac{\partial^2 {u_1}^*}{\partial {x_2}^{*2}} \tag{8.15}$$

From our discussion of the characteristics of the boundary layer we have the following boundary conditions:

(i) $x_2 = 0 : u_1 = 0, u_2 = 0$ (no-slip),

(ii) $x_2 = \delta : u_1$ = U

(iii) $x_1 = 0 : u_1$ = U

This equation seems to have eliminated all parameters (like viscosity, density and Re). However, recall that x_2 is non-dimensionalized using δ and also u_2 is non-dimensionalized using δ. In order to determine δ one needs to know the relationship between δ and Re. So even though we may be able to solve Eqn. (8.15) there is still work to do to interpret the results.

Laminar Boundary Layer Solution

The solution to the boundary layer equations for steady flow over a flat surface parallel with the oncoming flow, with the associated boundary conditions, is called the Blasius solution. This was accomplished around 1913 originally by Paul Blasius, a graduate student of Prandtl's. Noting that Eqn. (8.15) shows that the basic equation consists of two convective acceleration terms and one viscous term which can be written in terms of the original dimensional variables as:

$$\rho\left(u_1\frac{\partial u_1}{\partial x_1}+u_2\frac{\partial u_1}{\partial x_2}\right)=\mu\left(\frac{\partial^2 u_1}{\partial x_2}\right) \tag{8.16}$$

with the conservation of mass as Equation (8.3).

Following the general solution approach of Blasius we can introduce the streamfunction, and use it to replace the velocity components:

$$u=\frac{\partial\psi}{\partial x_2};\quad v=-\frac{\partial\psi}{\partial x_1} \tag{8.17}$$

We have previously shown that this definition for two dimensional, incompressible flow satisfies conservation of mass, by inserting these expressions into Equation (8.3). Substituting these two expressions in terms of ψ for the two velocity components in Equation (8.16) results in:

$$\frac{\partial\psi}{\partial x_2}\left(\frac{\partial^2\psi}{\partial x_1\partial x_2}\right)-\frac{\partial\psi}{\partial x_1}\left(\frac{\partial^2\psi}{\partial x_2^2}\right)=\nu\left(\frac{\partial^3\psi}{\partial x_2^3}\right) \tag{8.18}$$

Here we have also divided through by density to obtain the kinematic viscosity, $\nu=\mu/\rho$, as the coefficient of the viscous term which is the only parameter in the equation (a constant for a given fluid). Although the resulting equation seems formidable the use of the streamfunction has reduced the two velocities into a single dependent variable ψ.

The streamfunction, with SI units of m^2/s, is nondimensionalized as:

$$f(\eta)=\frac{\psi}{\nu x_1\frac{1}{2}} \tag{8.19}$$

a non-dimensional similarity variable,η, is defined to combine x_1 and x into a varible:

$$\eta = \frac{x_2}{\left(\frac{\nu x_1}{U}\right)^{1/2}} = x_2\left(\frac{U}{\nu x_1}\right)^{1/2} \tag{8.20}$$

The new variable f is assumed to be only a function of η. This will be verified by inserting these new variables into the original equation and see if the result is only η dependent. Each of the terms in Eqn. (8.18) are transformed into the (f, η) variables using the chain rule similar to what was done in Chapter 7 to determine the governing equation for a suddenly accelerated flat surface. Similarly, the non-dimensional stream function of Eqn. (8.18) can be written in terms of u_1 and u_2 and inserted into Eqn. (8.16). The result is the following, using the prime designation to represent the derivative with respect to η

$$u_1 = \frac{\partial\psi}{\partial x_2}\frac{\partial\psi}{\partial\eta}\frac{\partial\eta}{\partial x_2} = \frac{\partial\psi}{\partial\eta}\left(\frac{U}{\nu x_1}\right) = f'\left(\frac{U}{\nu x_1}\right) \tag{8.21}$$

$$u_2 = -\frac{\partial\psi}{\partial x_1} = \left(\frac{vU_\infty}{4x}\right)^{1/2}(\eta f' - f) \tag{8.22}$$

$$\frac{\partial u_1}{\partial x_1} = Uf''\frac{\partial\eta}{\partial x_2} = -\frac{uf''\eta}{2x_1} \tag{8.23}$$

$$\frac{\partial u_1}{\partial x_2}Uf''\frac{\partial\eta}{\partial x_2} = Uf''\left(\frac{U}{\nu x_1}\right)^{1/2} \tag{8.24}$$

$$\frac{\partial^2 u_1}{\partial {x_2}^2} = \frac{U^2 f'''}{\nu x_1} \tag{8.25}$$

Substituting all of this into our boundary layer equation, (8.16), with a little bit of manipulation the result becomes:

$$f''' + \frac{1}{2}ff'' = 0 \tag{8.26}$$

This now is a single variable ordinary differential equation, which shows that $f = f(\eta)$ resulting in a similarity equation. To obtain a solution the boundary conditions need to be specified in terms of η. Specifically there are three: $x_2 = 0 : u_1 = u_2 = 0$ (no slip condition) so

$$f(0) = f'(0) = 0$$

$x_2 \geq \delta : u_1 = U$ so

$$f'(\eta \to \infty) = 1.0 \tag{8.27}$$

The no slip condition for u_2 results in $f(0) = 0$ since $f'(0) = 0$ for the no slip condition of u_1 so by Eqn. (8.22) $f(0) = 0$. Also, we take the boundary condition at the edge of the boundary as $\delta \to \infty$ since U remains constant for $x_2 > \delta$. Physically this imposes an asymptotic limit of the solution for large δ, or beyond the edge of the boudary layer.

The solution to Equation (8.26) along with the boundary conditions of Equations (8.27) is rather straight forward using a Runge-Kutta technique. However, this was not the case when Blasius solved this equation numerical by hand back in the early 1900s at the University of Göttingen.

The results of the numerical solution of Eqn (8.26) are expressed as $f'(\eta) = u_1/U_\infty$ and are shown in Figure (8.3). Obviously u_1/U varies from zero at the surface ($\eta = 0$) and approaches 1.0 at $x_2 = \delta$. The value of δ can be determined from this plot, however one must select a value of $f'(\delta)$ close to 1.0, and typically 0.99 is chosen. At this value of $f'(\delta)$ the value of $\eta \simeq 5.0$. Using the definition of η given above and setting it equal to 5.0 when $x_2 = \delta$ results in:

$$5 = \frac{\delta}{\left(\frac{\nu x_1}{U}\right)^{1/2}}$$

Rearranging this yields an expression for the boundary layer versus x_1:

$$\frac{\delta}{x_1} = \frac{5}{{Re_{x_1}}^{1/2}} \tag{8.28}$$

This shows that $C = 5$ in Eqn. (8.13). Eqn. (8.28) allows for the evaluation of the boundary layer thickness, δ, at any x_1 location for a given value of U and viscosity ν. Using the solution in Fig. (8.3) also allows for the evaluation of the velocity components, u_1 and u_2 and any location x_1 and x_2 in the boundary layer.

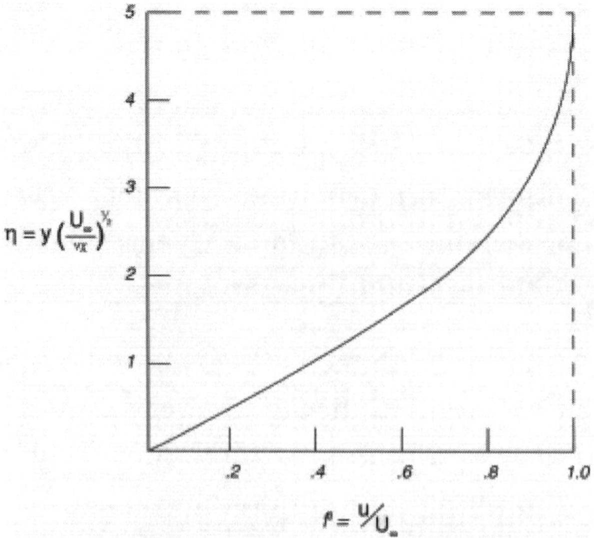

Fig 8.3 Blasius solution to steady laminar flow over a flat plate.

Other aspects of the flow can also be found from the solution. The total force on the surface caused by fluid flow friction can be found from the surface stress, $\tau_s = \mu \frac{\partial u_1}{\partial y}$, which requires evaluating the velocity derivative at the surface, $x_2 = 0$. The non-dimensional form is used to evaluate $\frac{\partial u_1}{\partial y}$ at $x_2 = 0$. Using Eqn. (8.24) we have:

$$\left(\frac{\partial u_1}{\partial x_2}\right)_{x_2=0} = f''(0)U^{3/2}\sqrt{\frac{\rho}{\mu x_1}} \tag{8.29}$$

where $f''(0)$ is the second derivative of f evaluated at $\eta = 0$. From the Blasius solution $f''(0) = 0.332$. After multiplying Equation (8.29) by the dynamic viscosity to obtain the surface stress and then dividing by $\frac{1}{2}\rho U^2$ which defines the surface skin friction coefficient, c_f, the result is:

$$c_f = \frac{\tau_s}{\frac{1}{2}\rho U^2} = \frac{0.664}{Re_{x_1}{}^{1/2}} \tag{8.30}$$

The above expression can be used to find the local surface stress at any x_1 location along the surface (recall $Re_{x_1} = \frac{\rho U_\infty x_1}{\mu}$). This then indicates that for a given freestream velocity of a given fluid the surface shear stress decreases as $x_1{}^{-1/2}$. The final step to determine the total drag force, F_D, on the surface is to integrate τ_s over the entire surface using the results of Equation (8.30). $F_d = \int_0^L \tau_s dx_1 S$ which is written as the drag coefficient, C_D:

$$C_D = \frac{F_D}{\frac{1}{2}\rho U^2 (LS)^2} = \frac{1.328}{{Re_L}^{1/2}} \tag{8.31}$$

Where S is the span of the surface in the x_3 direction.

The drag coefficient provides a nondimensional convenient expression to determine the overall drag force on a surface. Knowing the length of plate, L, one calculates the Re_L and then uses (8.31) to find F_D.

Pressure Gradient Effects

The above discussion is concerned with forces caused by fluid friction for flow over a surface. The other major contribution to the total force acting on a surface when there is flow over the surface is that of pressure. When the pressure is uniform within the flow it has no contribution to the change of momentum of the flow (zero pressure gradient) along the flow. However, as we have seen for inviscid flow, a pressure distribution over a surface results in a force and this pressure distribution is significantly altered by the flow conditions. Consider flow over a cylinder or Rankine oval where the velocity variation results in a pressure variation at the surface of the oval. The force for inviscid flow can be obtained from the Bernoulli equation to evaluate the local pressure which is then integrated over the surface.

For viscous flows the pressure and surface frictional forces both need to be integrated over the surface. However, importantly, these two can not be treated as independent of each other. That is to say, the pressure distribution is influenced by the velocity distribution, or vice versa, as we see from the Navier-Stokes equation which contains a pressure gradient term. Consequently, one can expect that the velocity distribution within the boundary layer will be altered as the pressure gradient changes (in other words we no longer have Eqn. (8.3) which assumes $\frac{\partial P}{\partial x_1} = 0$). With an altered velocity profile the velocity gradient at the surface also changes indicating that the surface stress changes. The link between pressure gradient and surface stress can be very complex, influenced by the conditions that result in a specific pressure gradient. This is made apparent by considering the Navier-Stokes equation evaluated at the surface, with no slip boundary conditions. In this case acceleration terms are all zero and there is a balance of pressure and viscous forces at the surface:

$$\frac{\partial P'}{\partial x_1} = \mu \left(\frac{\partial^2 u_1}{\partial {x_2}^2} \right)_{x_2=0} = \left(\frac{\partial \tau}{\partial x_2} \right)_{x_2=0} \tag{8.32}$$

For the boundary layer, assuming it is thin, the pressure gradient can be evaluated outside the boundary layer as discussed previously. Also, in this expression we have absorbed the body force term into the pressure term: $P' = P + \rho g h$ where h is the height above some datum. This form of the pressure plus body force accounts for the hydrostatic pressure variation that may occur along the direction x_1 caused by changes in elevation given by changes in h.

Eqn. (8.32) provides insight into the flow conditions very close to the surface for changing pressure gradients. For instance, if $\frac{\partial P'}{\partial x_1} < 0$ (a "*favorable pressure gradient*") then the pressure decreases in the flow (or x_1) direction. This implies from Eqn (8.32) that the surface stress decreases moving away from the surface as one might expect since it ultimately reaches zero at the edge of the boundary layer, for inviscid flow conditions. For conditions of very large favorable pressure gradients ($\frac{\partial P'}{\partial x_1} \ll 0$) the decrease of stress with x_2 is greater and the boundary layer becomes thinner, and the wall shear stress becomes larger. The distribution of stress is shown in qualitatively Fig. 8.4.

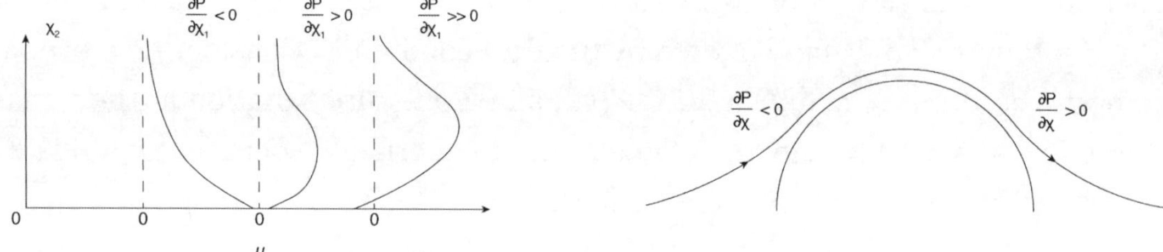

Fig. 8.4 Stress distribution for favorable and adverse pressure gradients in boundary layer flows; flow over a cylinder illustrating favorable and adverse pressure gradients for flow over a cylinder.

If the pressure gradient has a positive sign, $\frac{\partial P'}{\partial x_1} > 0$, (an "*adverse pressure gradient*") the stress increases with increasing x_2 away from the surface again as shown in Eqn. (8.32). This implies that at some point the stress reaches a maximum, and then decays to zero at the edge of the boundary layer. See Fig. 8.4. As the adverse pressure gradient increases one can suspect that the gradient of stress near the surface also becomes larger, forcing the stress to approach zero or even go negative as shown in Fig. 8.4. Under these conditions, of negative stress, the implications are that the sign reversal causes the flow in a region near the wall to actually go in the negative, or upstream, x_1 direction. This is the definition of "*flow separation*". Separated flow no longer adheres to boundary layer flow as part of the flow near the surface is moving upstream and part of the flow further from the surface is moving downstream. This generally results in swirling flow near the surface.

What can cause such a large adverse pressure gradient? One common situation is the existence of large surface curvature, such as flow over a cylinder of diameter D and length S into the page, shown in Fig. 8.5. In these cases the x_1 coordinate is taken as curving along the surface in the flow direction. On the upstream side of the cylinder the pressure gradient is negative as the flow accelerates with increasing velocity. The growth of δ is not as large as for the zero pressure gradient case, as noted above. As the flow continues over the top (or bottom) of the cylinder the pressure gradient becomes positive, increasing pressure in the flow direction. This is caused by the decrease of velocity for the flow beyond the top (bottom) of the cylinder. The inviscid flow over a cylinder has been studied previously (as a superposition of uniform flow and doublet). For this inviscid case the pressure can be found from the Bernoulli equation and the result can be expressed as the pressure coefficient as:

$$c_p = \frac{P - P_\infty}{\frac{1}{2}\rho U^2} = 1 - \frac{{v_s}^2}{U^2} = 1 - 4sin^2\theta \tag{8.33}$$

where P_∞ is the pressure for upstream (or downstream) of the cylinder and v_s is the surface velocity at any point around the cylinder (zero at the front and back stagnation points and maximum at the top and bottom). A plot of this inviscid pressure distribution around the cylinder is shown in Fig. 8.5. Notice that the pressure distribution is symmetric and the net force caused by pressure is zero. The maximum pressure is at the stagnation points, as expected, and equal to $P_\infty + \frac{1}{2}pU^2$ from Bernoulli equation.

With the addition of viscous effects, for flow over a cylinder, as described above, the adverse pressure gradient on the rear portion of the cylinder results in flow separation when the freestream flow is sufficiently large. The flow then forms a "wake" across the back of the cylinder consisting of swirling, turbulent flow. The result is an asymmetric flow pattern around the cylinder and an asymmetric pressure distribution. In the wake the pressure is not as high as in the inviscid case. We see in the pressure coefficient plot that for viscous flows the pressure recovery on the downstream side is not as large as the inviscid case and the net affect is a downstream drag force. Turbulent flows tend to have a greater pressure recovery and a lower nondimensional drag coefficient ($C_D = \frac{F_D}{\frac{1}{2}\rho U^2 DS}$) when compared with laminar flow.

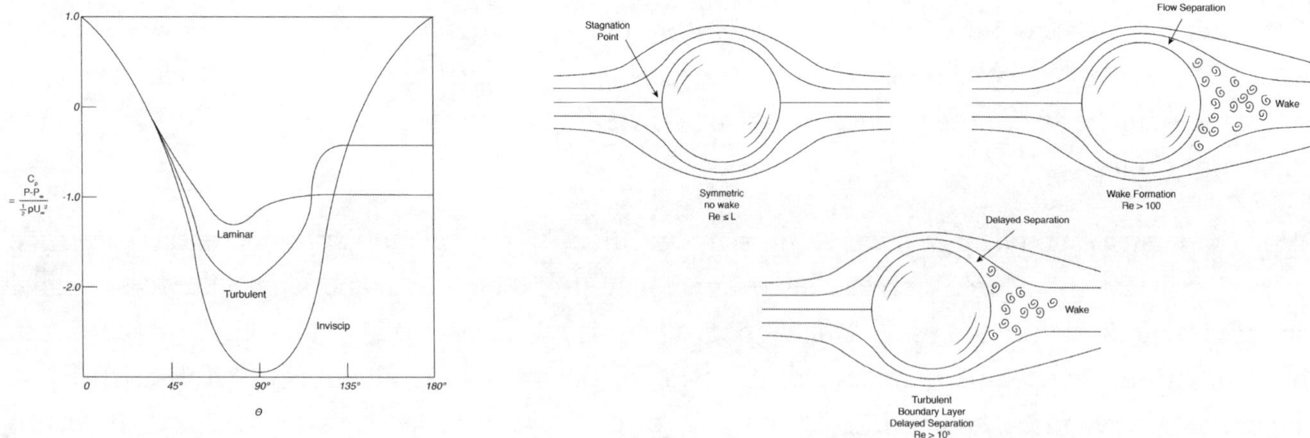

Fig. 8.5 Pressure distribution for flow over a cylinder and illustration of wake formation for laminar and turbulent flow.

IX. Integral Boundary Layer Relationships

Historically, the development of the integral form of the boundary layer equations, as is presented here, has provided a powerful tool to evaluate surface viscous forces for more complicated flows with pressure gradients. In the Blasius solution presented in the last chapter, the velocity profile is determined directly from a modified form of the Navier-Stokes equation. From this solution, the local velocity derivative at the surface, and therefore the surface shear stress, can be found as a function of position along the surface. Using the definition of the surface stress the total force along the surface can then be found. The integral method in many ways takes the opposite approach. The basic governing equations are integrated across the flow using the wall shear stress as an unknown boundary condition, eventually to be solved for. The local pressure gradient, which may vary in the freestream velocity direction, is taken to be a parameter known at each position along the flow. This can be done since the assumption is that the pressure gradient in the freestream is identical to the pressure gradient in the boundary layer. This is part of the boundary layer assumptions and is valid for both laminar and turbulent flows. In arriving at the integral form of the boundary layer equations there are several scaling parameters introduced along the way that are often used to characterize the boundary layer flow and help interpret the flow conditions. These parameters can be used to help scale the flows and predict surface force distributions.

Scaling Parameters

Boundary layer thickness

The boundary layer thickness, as we have seen, is given by δ, and it represents the thickness of the boundary layer measured from the surface to the location where the velocity reaches the freestream value, U. However, since the velocity within the boundary layer asymptotically approaches U it is often difficult to obtain an accurate measure of δ. To circumvent this problem the boundary layer thickness can be taken to be when the velocity reaches, say, 1% of U. As discussed in the last chapter, in a favorable pressure gradient where the pressure decreases along the flow direction,the boundary layer will be thinner when compared with a zero pressure gradient. The additional pressure force in the flow direction increases the local velocity at any given position and the boundary layer reaches the freestream value at a lower value of the cross stream coordinate. This then reduces the boundary layer thickness. The opposite is true for an adverse pressure gradient. Consequently, the boundary layer thickness is directly influenced by the pressure gradient and as such there is a similar impact on the surface stress. A thinner boundary layer caused by a favorable pressure gradient, with the same freestream velocity, will have a larger surface velocity gradient and a higher local surface shear stress.

In general we note that the boundary layer thickness is a function of the downstream coordinate, x_1, and δ increases in the x_1 direction. An alternative way of expressing this is through the Reynolds number, since the Reynolds number depends on x_1 $\left(Re_{x_1} = \frac{\rho U x_1}{v} \right)$:

$$\delta = f(Re_{x1})$$

Displacement thickness, δ_1

The displacement thickness, δ_1, and momentum thickness, δ_2, are often used as a measure of the thickness of the boundary layer but are not actually the boundary layer thickness; these two are important parameters in the analysis of the frictional forces on the surface. The displacement thickness is, in general, a function of position, x_1, along the flow direction as is δ. Typically δ_1 increases with increasing x_1 and is less than δ. A sketch illustrating the definition of the displacement thickness is given in Fig. 9.1 where δ_1 is determined based on the mass flow rate within the boundary layer. The velocity distribution can be integrated across the boundary layer, normal to the surface, to determine the mass flow rate at any location, x_1 as

$$\dot{m} = \int_0^\delta \rho u_1 dx_2 \; S \tag{9.1}$$

where u_1 is the x_1 component of the velocity and S is the span, or the distance into the page. The distribution of u_1 follows the two boundary conditions, $u_1 = 0$ at the surface and $u_1 = U$ at the edge of the boundary layer, δ.

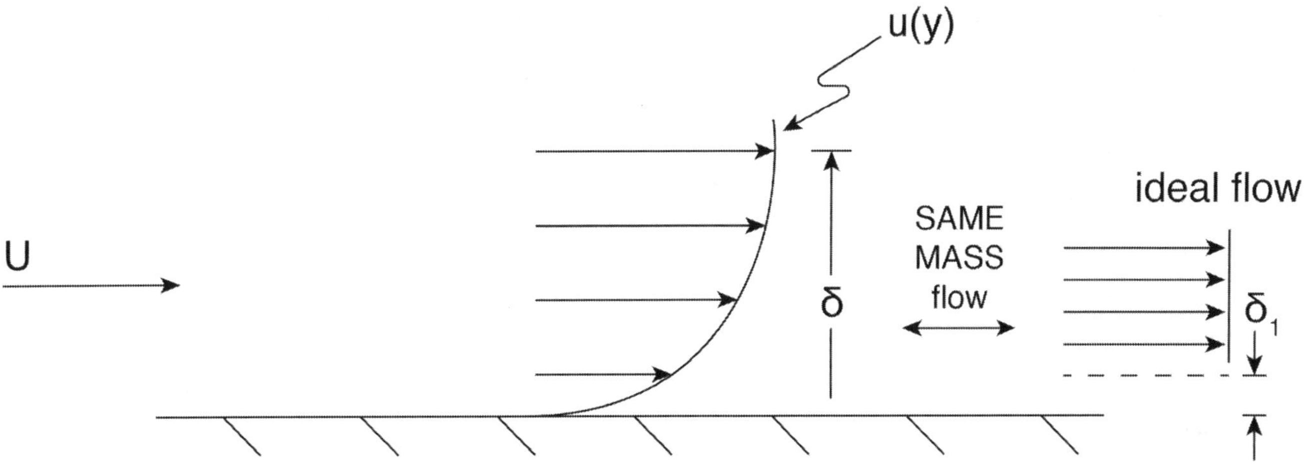

Fig. 9.1 Illustration of the definition of the displacement thickness based on the total mass flow rate within the boundary layer; note that the mass flow rate changes along the flow in the x_1 direction consequently δ_1 is a function of x_1.

The surface generated friction by the surface on the flow causes the velocity to decrease at any given x_2 position along the surface. This decrease across the entire boundary layer results in a decreasing flow rate within the boundary layer with increasing x_1. Therefore the amount of mass flow lost in the boundary layer due to surface friction can be thought of as local a mass flow deficit. This deficit can be calculated by comparing the actual velocity distribution to the ideal velocity without friction, which would be U across the entire boundary layer. The local mass flow deficit at any x_2 location is then $d\dot{m}_{deficit} = \rho\,(U - u)\,dy\;S$ and the total mass flow deficit is obtained by integrating this across the flow as:

$$\int_0^{\delta} d\dot{m}_{deficit} = \dot{m}_{deficit} = \int_0^{\delta} \rho\,(U - u_1)\,dy\;S \tag{9.2}$$

The mass flow deficit can be equated to mass flow rate loss given as the product of density, U and (δ_1 S), where δ_1 is an unknown "thickness". So setting the integral term in Eqn. (9.2) to this product $(\rho U \delta_1 S)$ and solving for δ_1 results in:

$$\delta_1 = \int_0^{\delta} \left(1 - \frac{u_1}{U}\right) dy \tag{9.3}$$

which assumes that the density is constant. Eqn. (9.3) is the definition of the displacement thickness and can be interpreted as follows. If the total mass flow rate lost in the boundary layer due to friction is divided by $\rho U S$ the result is a length that would represent a distance normal to the surface through which all of the lost mass flow rate would pass if it traveled at velocity U(the velocity it would have if there were no friction.) Another way of looking at this distance is that it represents the vertical distance the surface would need to be moved upward (in the x_2 direction) to capture all of the lost mass flow rate if the velocity were uniformly at U, or frictionless. Either of these interpretations indicates that as δ_1 grows larger along the flow there is an increase in the total mass fow deficit. Since surface friction slows more and more fluid, creating a larger boundary layer thickness in the x direction, then δ_1 must increase in the downstream x_1 direction. At any given position along the surface the greater the local surface shear stress the greater the value of δ_1. Consequently, a favorable pressure gradient which increases the local shear stress compared with zero pressure gradient, results in a larger relative value of displacement thickness.

Momentum thickness, δ_2

The momentum thickness, δ_2, has somewhat of the same interpretation as δ_1, but it has to do with the momentum rate loss by the frictional forces, rather than the mass flow rate loss. In this case we develop a measure for the momentum rate loss, or momentum deficit, generated by friction. Noting that the momentum rate lost crossing a given vertical plan across the boundary layer is given by:

$$\int_0^\delta \rho u \left(U - u_1\right) dx_2 \ S \tag{9.4}$$

where $\rho u dx_2 \ S$ is the mass flow rate within an elemental area $dx_2 S$ w and the expression $(U - u_1)$ is the momentum lost per mass flow rate. The product of these two terms integrated over the entire boundary layer δ is the momentum loss rate at any given x_1 position along the surface. Similar to what was done for the mass flow rate deficit, here we imagine that the momentum rate lost due to friction is equal to the mass flow rate through an area equal to $(\delta_2 S)$ with velocity U or

$$\rho U \delta_2 S U = \rho U^2 \delta_2 S = \int_0^\delta \rho u \left(U - u_1\right) dx_2 \ S$$

Dividing through by $\rho U^2 S$ yields the definition of the displacement thickness:

$$\delta_2 = \int_0^\delta \frac{u}{U}\left(1 - \frac{u_1}{U_\infty}\right) dx_2 \tag{9.5}$$

We can interpret this as the distance δ_2 the surface would need to be moved in the x_2 direction into the flow if the velocity is uniformly at U in order to reduce the momentum rate equivalent to the actual rate of momentum loss. We can also interpret this somewhat differently as follows.

A flat surface is shown in Fig. 9.2 indicating the surface shear stress, τ_s, acting on the fluid. A control volume with an elemental length along the surface, x_1, and extending upward to the edge of the boundary layer is shown in the figure. The sum of all of the forces acting on the fluid in the control volume consists of τ_s, since there is zero pressure gradient along the flow, and we assume there is no frictional forces at the top edge of the boundary layer, by definition of the extent of the boundary layer. The momentum equation for this control volume becomes the sum of the forces equal to the change of momentum rate leaving compared to that entering:

$$-\tau_s dx_1 S = \left(\rho U^2 \delta_2 S \right)_{x+dx_1} - \left(\rho U^2 \delta_2 S \right)_{dx_1} = \left(\rho U^2 d\delta_2 S \right) \tag{9.6}$$

where $d\delta_2$ is the change of the displacement thickness along dx_1, while noting that $\rho U^2 S$ is constant along x_1. The negative sign is included to indicate that the stress on the fluid is in the negative x_1 direction at the surface. If we rearranging this equation and change the stress to be that on the surface caused by the fluid flow (change the sign of τ_s) then:

$$\tau_s = \rho U^2 \frac{d\delta_2}{dx_1} \tag{9.7}$$

Note that this equation is valid for any particular velocity distribution that may exist in the boundary layer. Also, it indicates that the rate of change of the momentum thickness decreases along the flow direction since the surface stress decreases along the flow direction. If one can determine $\frac{d\delta_2}{dx_1}$ it is possible to find the wall surface stress, τ_s.

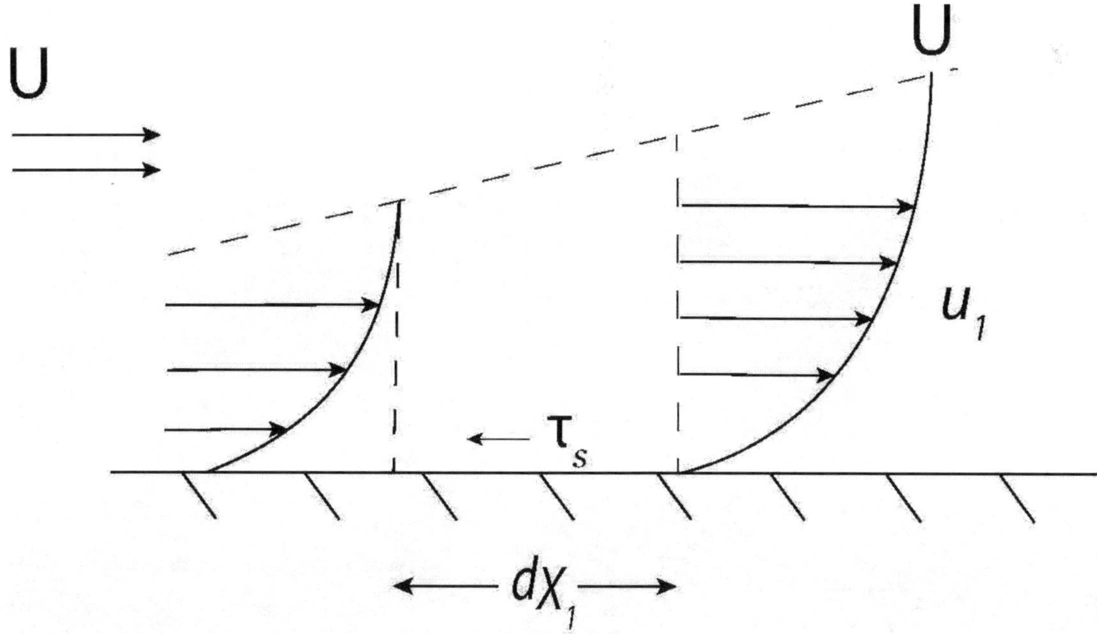

General Integral Boundary Layer Equations

The general integral boundary layer equation is obtained by starting with the Navier- Stokes equation simplified for boundary layer flow:

$$u_1\frac{\partial u_1}{\partial x_1}+u_2\frac{\partial u_1}{\partial x_2}=-\frac{1}{\rho}\frac{\partial P}{\partial x_1}+\nu\frac{\partial^2 u_1}{\partial x_2} \tag{9.8}$$

Note that $\nu=\frac{\mu}{\rho}$, and is the kinematic viscosity. Since the pressure gradient term can be determined from the freestream flow which is assumed to be inviscid the Euler equation can be used where we ignore gravitational effects such that:

$$-\frac{1}{\rho}\frac{\partial P}{\partial x_1}=U\frac{dU}{dx_1} \tag{9.9}$$

Note that body force terms could be included by replacing P with $P'=P+\rho gh$.

Equation (9.8) is integrated across the boundary layer, in the x_2 direction, for a given value of x_1. The boundary conditions for this are no-slip at x_2 = 0, and u = U at $x_2=\delta$. Before we complete this operation we examine the second term on the left hand side in Eqn. (9.8). First we use continuity for a two dimensional flow:

$$\frac{\partial u_2}{\partial x_2}=-\frac{\partial u_1}{\partial x_1} \tag{9.10}$$

which integrates to:

$$u_2=-\int_0^{\delta}\frac{\partial u_1}{\partial x_1}dx_2 \tag{9.11}$$

and this expression is used in the second term of (9.8). Next each term in Eqn. (9.8) is integrated from $y=0$ to $y=\delta$. For instance the second term becomes:

$$\int_0^{\delta}u_2\frac{\partial u_1}{\partial x_2}dx_2=-\int_0^{\delta}\frac{\partial u_1}{\partial x_2}\left(\int_0^{x_2}\frac{\partial u_1}{\partial x_1}dx_2\right)dx_2 \tag{9.12}$$

This term is now integrated by parts:

$$\int_1^2 b\,da=ab\big|_1{}^2-\int_1^2 a\,db$$

where we set

$$db = \frac{\partial u_1}{\partial x_1}dx_2, da = \frac{\partial u_1}{\partial x_2}dx_2 = \partial u_1 \text{ and } a = u_1$$

The result for this term using the boundary conditions for u_1 is:

$$-U\int_0^\delta \frac{\partial u_1}{\partial x_1}dx_2 + \int_0^\delta u_1\frac{\partial u_1}{\partial x_1}dx_2 \tag{9.13}$$

We insert (9.13) into (9.8) while integrating all other terms and obtain:

$$\int_0^\delta u_1\frac{\partial u_1}{\partial x_1}dx_2 - U\int_0^\delta \frac{\partial u_1}{\partial x_1}dx_2 + \int_0^\delta u_1\frac{\partial u_1}{\partial x_1}dx_2 = \int_0^\delta U\frac{dU}{dx_1} + \int_0^\delta \nu\frac{\partial^2 u_1}{\partial x_2}dx_2 \tag{9.14}$$

The last term is the viscous term and can be expressed in terms of the shear stress, $\tau = \mu\frac{\partial u_1}{\partial x_2}$ (the x_1 derivative of u_2 is taken to be small compared with the x_2 derivative of u_1, consistent with the boundary layer approximations.) So the last term is:

$$\int_0^\delta \nu\frac{\partial^2 u_1}{\partial x_2}dx_2 = \int_0^\delta \frac{1}{\rho}\frac{\partial\tau}{\partial x_2}dx_2 = \frac{1}{\rho}[\tau(x_2=\delta) - \tau(x_2=0)] = -\frac{\tau_s}{\rho} \tag{9.15}$$

We combine the first and third terms of Eqn. (9.14) as:

$$\int_0^\delta 2u_1\frac{\partial u_1}{\partial x_1}dx_2 = \int_0^\delta \frac{\partial(u_1u_1)}{\partial x_1}dx_2 \tag{9.16}$$

Then we rewrite the second term of Eqn. (9.14) by bringing U inside the integral since it is not a function of x_2, but may be a function of x_1:

$$-U\int_0^\delta \frac{\partial u_1}{\partial x_1}dx_2 = \int_0^\delta \left(-\frac{\partial(uU)}{\partial x_1} + u\frac{\partial U}{\partial x_1}\right)dx_2 \tag{9.17}$$

Getting closer to the final result we combine all these expressions in (9.14) to obtain:

$$\int_0^\delta (\frac{\partial(u_1u_1)}{\partial x_1} - \frac{\partial(u_1U)}{\partial x_1} + u_1\frac{dU}{dx_1} - U\frac{dU}{dx_1})dx_2 = \frac{\tau_s}{\rho} \tag{9.18}$$

This is rearrange as:

$$\int_0^\delta \frac{\partial\left(u_1\left(u_1-U\right)\right.}{\partial x_1} dx_2+\int_0^\delta \frac{dU}{dx_1}\left(u_1-U\right) dx_2=-\frac{\tau_s}{\rho} \tag{9.19}$$

Exchanging the order of integration and derivatives in the first term, changing sign of all the terms, yields:

$$\frac{\partial}{\partial x_1} \int_0^\delta u_1\left(U-u_1\right) dx_2+\frac{dU}{dx_1} \int_0^\delta\left(U-u_1\right) dx_2=\frac{\tau_s}{\rho} \tag{9.20}$$

This now allows us to use the definitions of δ_1 and δ_2 to arrive at our final form as:

$$\frac{d}{dx_1}\left(U^2 \delta_2\right)+\delta_1 U \frac{dU}{dx_1}=\frac{\tau_s}{\rho} \tag{9.21}$$

We end up with a differential equation in terms of the variables δ_1 and δ_2 both of which are functions of x_1. Recall that U(x_1) is determined by the pressure gradient through use of Eqn. (9.9), Euler's equation. When U = constant, as in flow over a flat plate this equation can be simplified to:

$$\frac{d\delta_2}{dx_1}=\frac{\tau_s}{\rho U^2} \text{ or } 2 \frac{d\delta_2}{dx_1}=\frac{\tau_s}{1/2 \rho U^2}=c_f \tag{9.22}$$

where the right hand side is the skin friction coefficient.

The problem of determining the local surface stress distribution is now reduced to Eqn. (9.21) for flows over surfaces with a pressure gradient or (9.22) for a flat plate with no pressure gradient. Next, we will examine methods to arrive at solutions to this.

Solution Method

There are several methods available to arrive at a solution for Eqn. (9.21). We present a typical method that is based on the assumption of a velocity profile in nondimensional form across the flow at any given position, x_1. The key to this approach is based on the fact that the Blasius solution found a similarity solution and collapsed the velocity profile onto one curve when using proper scaling of the velocity and cross-stream coordinate, the latter also incorporated the stream-wise coordinate to help scale it properly. Here we will assume a non-dimensional velocity profile and use it to determine the integrals in the definitions of δ_2 and δ_1.

The proper scaling of the cross-stream coordinate is taken to be the boundary layer thickness, where δ is a function of x_1 then this scaling combines the cross-stream and stream-wise dependency. We define this new non-dimensional variable as:

$$\eta = \frac{x_2}{\delta} \tag{9.23}$$

being careful to note that this variable, η, is not the same as used in the Blasius definition. Next we assume that the nondimensional velocity, $f = \frac{u_1}{U}$, is some function of η or:

$$f = f(\eta) \tag{9.24}$$

One may say that it is necessary to precisely determine this function since the surface stress, τ_s, depends on the velocity derivative at the surface. This certainly is the case for the Blasius approach. However, examining Eqns. (9.21) and (9.22) it can be seen that the surface stress is a function of the integral of the velocity distribution as expressed through the variables δ_1 and δ_2 . Based on this we conclude that to get a solution we must obtain good values for the integrals expressed in these equations, rather than the surface derivative of the velocity.

Taking this approach it is possible to assume a velocity distribution using the assumed nondimensional form in Eqn. (9.24) requiring it to satisfy the boundary conditions. One can use this distribution after transforming the integrals in (9.21) into the nondimensional form. We will use the example of flow over a flat plate, governing by Eqn. (9.22). This simplifies the details, but the same approach is used for flows with pressure gradients, but requires further input of a known pressure gradient.

First looking at Eqn. (9.22) written in terms of the skin friction coefficient we have:

$$c_f = 2\frac{d\delta_2}{dx_1} \tag{9.25}$$

We can not evaluate δ_2 since we do not know the value of δ. We illustrate the process by assuming a polynomial velocity distribution based on Eqn. (9.24) variables:

$$f = a + b\eta + c\eta^2 + d\eta^3 + \text{higher order terms} \tag{9.26}$$

There are four parameters in this polynomial, a, b, c and d that need to be determined based on boundary conditions on the velocity distribution. Notice that the higher the order of the polynomial the more conditions are needed to evaluate these coefficients. The boundary conditions written in terms of the variables, f, δ are:

$$f(0) = 0 \text{ (no slip boundary condition at the surface)}$$

$$f(1) = 1 \text{ (velocity becomes the freestream value at } \delta)$$

$$\frac{\partial f}{\partial \eta} = 0 \text{ (no shear at } \delta)$$

$$\frac{\partial^2 f}{\partial \eta^2} = 0 \text{ (no net viscous force at the surface)}$$

The last condition is a consequence of all terms in the Navier-Stokes equation becoming zero at the surface if the pressure gradient is zero, and $u_1 = u_2 = 0$, and then transforming the viscous term into the non-dimensional variables. These four conditions can now be used to determine the four coefficients in Eqn. (9.26). The results are:

$$a = 0, \;\; b = 3/2, \;\; c = 0, \;\; d = -1/2$$

Inserting these into Eqn. (9.26) results in the velocity distribution:

$$f = \frac{3}{2}\eta - \frac{1}{2}\eta^3 \tag{9.27}$$

This result is unique to the selection of a third order polynomial for the velocity distribution. Next, we rewrite the relationship of c_f in terms of the nondimensional variables.

First we define the momentum and displacement thicknesses in terms of δ as:

$$\frac{\delta_1}{\delta} = \int_0^1 \left(1 - \frac{u_1}{U}\right) d\eta \tag{9.28}$$

$$\frac{\delta_2}{\delta} = \int_0^1 \frac{u_1}{U}\left(1 - \frac{u_1}{U}\right) d\eta \tag{9.29}$$

Here we have changed the integration from dx_2 to $d\eta$, where $dx_2 = \delta d\eta$ and need to change the limits of the integral accordingly. Notice that once we have an expression for $f = \frac{u_1}{U}$ from our polynomial, Eqn. (9.26), then both $\frac{\delta_1}{\delta}$ and $\frac{\delta_2}{\delta}$ are just constants. The important point now is that we can rewrite Eqn. (9.25) as:

$$c_f = 2\frac{d\delta_2}{dx_1} = 2\frac{d}{dx_1}\left(\left(\frac{\delta_2}{\delta}\right)\delta\right) = 2\left(\frac{\delta_2}{\delta}\right)\frac{d\delta}{dx_1} \tag{9.30}$$

So we have replaced having to know $\delta_2 = f(x_1)$ to needing to know $\delta = f(x_1)$. We find the latter relationship using our function $f(\eta)$ as:

$$\tau_s = \mu \frac{\partial u_1}{\partial x_2} = \mu \frac{\partial u_1}{\partial \eta} \frac{\partial \eta}{\partial x_2} = \frac{\mu U}{\delta} \frac{\partial f}{\partial \eta} = \frac{\mu U}{\delta} f'(0)$$

$$c_f = \frac{\tau_s}{\frac{1}{2}\rho U^2} = 2\frac{\delta_2}{\delta} \frac{d\delta}{dx_1} = \frac{\mu U}{\delta \frac{1}{2}\rho U^2} f'(0) \quad (9.31)$$

This last relationship is a differential equation for $\delta = f(x_1)$ since $\frac{\delta_2}{\delta}$ and $f'(0)$ are both constants known from Eqn. (9.26) using the boundary conditions to obtain Eqn. (9.27). Notice that these two constants are dependent on the order of the polynomial selected. Now Eqn. (9.31) can be integrated to obtain:

$$\frac{\delta}{x_1} = \frac{\left(2f'(0)\frac{\delta}{\delta_2}\right)^{1/2}}{{Re_{x_1}}^{1/2}} = \frac{C}{{Re_{x_1}}^{1/2}} \quad (9.32)$$

where $C = \left(2f'(0)\frac{\delta}{\delta_2}\right)^{1/2}$ and is a constant dependent on the order of the polynomial selected. Knowing $\delta = f(x_1)$ we can find $\frac{d\delta}{dx_1}$ and then c_f using Eqn. (9.31). The result is:

$$c_f = \frac{\delta_2}{\delta} \frac{C}{{Re_{x_1}}^{1/2}} \quad (9.33)$$

For our velocity profile, Eqn. (9.27) we have:

$$\frac{\delta_1}{\delta} = \frac{3}{8}, \frac{\delta_2}{8} = \frac{117}{840}, \text{and } f(0) = \frac{3}{2}.$$

This results in $C = 4.64$ and:

$$C_f = \frac{0.6464}{Re_{x_1}\frac{1}{2}} \quad (9.34)$$

The overall solution procedure now becomes pretty straight forward: (i) assume a functional form for $f(\delta)$, (ii) match required boundary conditions to find the coefficients in the equation, (iii) calculate the constant $\frac{\delta_2}{\delta}$, (iv) calculate the constant $f'(0)$, (v) finally calculate c_f, which determines the local value of the skin friction coefficient.

To determine the total friction drag force, F_D, on a surface of length L and area A_s = LS one needs to integrate the local surface stress along the distance L. The result is typically written in terms of the overall friction drag coefficient, C_f as:

$$C_f = \frac{F_D}{\frac{1}{2}\rho U^2 A_s} = \frac{1}{\frac{1}{2}\rho U^2 A_s}\int_0^L \tau_s dx_1 w = \frac{1}{L}\int_0^L c_f dx_1 = 2c_f(x_1 = L) \qquad (9.35)$$

where the last designation, $2c_f(x_1 = L)$, means two times the value of the skin friction coefficient evaluated at $x_1 = L$.

The result assuming the third order polynomial for the non-dimensional velocity profile given above is:

$$C_f = \frac{1.2928}{{Re_L}^{1/2}} \qquad (9.36)$$

The results when compared with experimental data for the appropriate Reynolds number conditions become very close to these predicted values. Consequently, this procedure is a fairly easy yet surprisingly accurate means to predict friction drag forces on flat surfaces. When pressure gradients come into play, such as surfaces with some degree of curvature it is more of a challenge to obtain a good velocity profile since for this case at the surface the net viscous forces are not zero but are balanced by the pressure gradient. The velocity profiles need to reflect this condition and there are a number of procedures available to address this. The interested reader is encouraged to see F.W. White's book, Viscous Fluid Flow for some examples.

X. Introduction to Turbulence Effects

The physical attributes of turbulent flow, in and of itself is a very difficult subject. That is to say, the flow characteristics of turbulence are difficult to measure, difficult to interpret and very difficult to incorporate into analyses. Never-the-less progress has been made, but much of it relies heavily on empirical results. Turbulence is typically described as a time varying phenomena of all dependent variables (except density for incompressible flows). In addition, spatial variations of the time dependency occurs as well. The results are that all of the velocity components and the pressure have spatio-temporal variations. Since the Navier-Stokes equations are nonlinear (through the convective acceleration terms) the statistical description of the flow relies on correlation functions among the various variables. There are a number of books written on the mathematical description of turbulence that can be challenging to incorporate into engineering applications. The nonlinear effects and some of the statistical measures of turbulence will become obvious once we look at the governing equations. For our purposes we will look at some of the physical consequences of turbulent flows and then illustrate how turbulent effects can be incorporated into an analysis of the viscous forces in boundary layer flows.

A few examples of turbulent flows are shown in Fig. 10.1. There are a wide range of fluctuations both in space and time. By scales we mean the magnitudes of local variations in time and space attributed to turbulent fluctuations about some local mean value. These are referred to as a broad spectrum of scales occurring within the flow. The flow geometry can have a large impact on the larger scales within the flow, since they interact often with the boundaries. However, the smaller scales tend to be mostly independent of the flow geometry indicating that there may be some underlying theoretical basis for their flow dynamics.

(a)

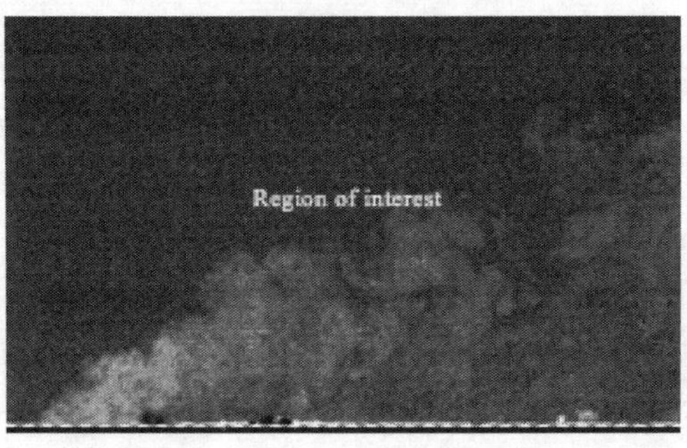

(b)

(c)

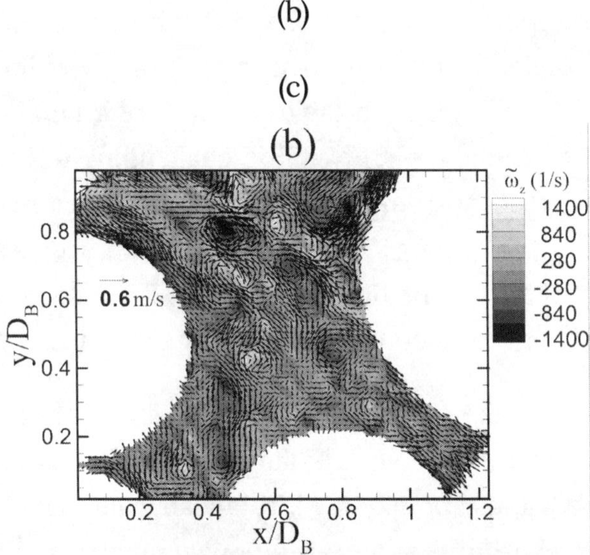

Fig. 10.1. Illustrations of turbulence flow visualization; (a) smoke visualization of flow over an inclined wing indicating flow separation from the leading edge which transitions to turbulence; (b) smoke visualization from a turbulent jet injecting flow into a cross flow in a wind tunnel indicating large and small scale fluctuations; (c) particle image velocimetry (PIV) of flow in a porous media (through a random packing of spherical beads) showing instantaneous velocity vector distribution indicative of small scale swirling motion.

Shown in Fig. 10.2 is a typical time history signal of a measured velocity component say from a hot wire anemometer at a given location within the flow. Although this signal may seem random in nature, it has been learned that there is a lot of structure to these kinds of signals. One thing to notice is that there are time events happening over many different time scales. There are small scale (or short) events with relative small fluctuating values coupled with longer time varying, higher amplitude fluctuations. The short time events seem to be superimposed over the longer events. To help illustrate this variation of time scales Fig. 10.2 also illustrates the "energy spectrum" of the time series signal. Here the mean square value of the fluctuations are decomposed by using a Fourier transformation of the signal. The mean square fluctuations have units of velocity squared that can be interpreted as the kinetic energy per mass of fluid. This plot identifies the frequency dependency of the energy associated with the signal. We see that the larger amplitude fluctuations occur at the longer time scales (lower frequency) and the smaller amplitude fluctuations occur at the shorter time scales (higher frequency). Turbulent flows are characterized by a broad spectrum within the signal typically with large amplitude (or energy) fluctuations having low frequency and low amplitude high frequency components. Looking at Fig. 10.2 we see that the longer time scales have several orders of magnitude greater energy than the smallest scales. This distribution has implications concerning the mixing characteristics of turbulent flows such as in the ability to promote high reaction rates in combustion and other chemical reactions. In contrast the energy spectrum of a flow that has a single oscillating frequency would have an energy spectrum that would appear as a spike at the frequency of oscillation and would not be referred to as turbulent.

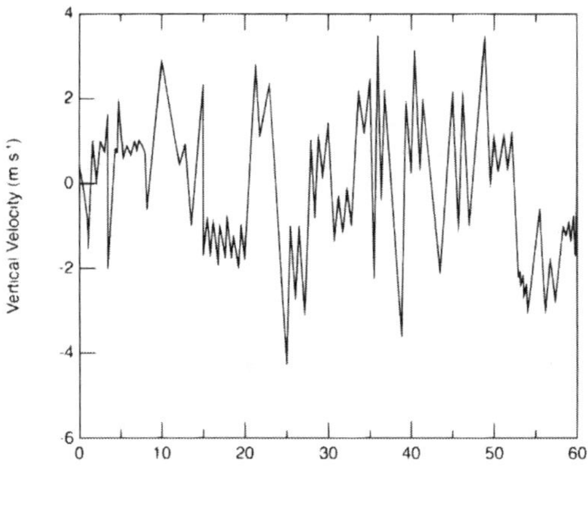

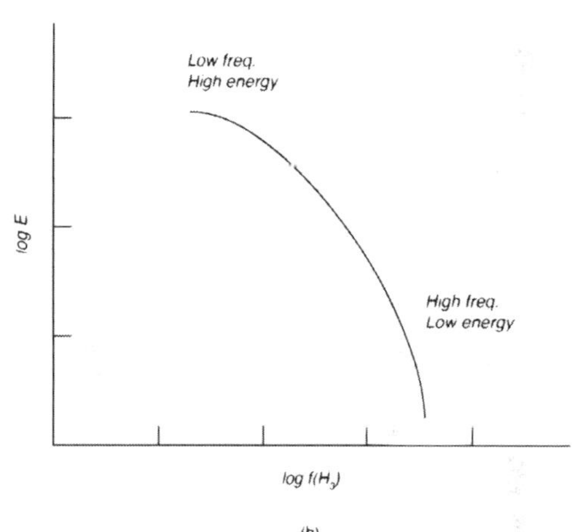

Fig. 10.2: (a) Typical velocity signal of a single component of velocity in turbulent flow measured as a given location, notice the large and small scale fluctuations that are superimposed on the signal. (b) energy spectrum on a log-log scale of the mean square velocity fluctuations verses frequency of fluctuations, notice the large range of energy and frequency that occurs.

The integration of the energy spectrum over all frequencies yields the total energy of the fluctuations in the signal. This is equivalent to the time average of the square of the fluctuations (the mean square value) $\overline{u'^2}$, where the fluctuating velocity is $u' = (u - \overline{u})$ with $\overline{u}$ being the time average value of the velocity and u is the instantaneous value of the velocity. This can be used to define the "turbulent intensity" as:

$$\frac{\overline{(u'^2)}^{1/2}}{\overline{u}} = \text{"turbulent intensity"}$$

Notice that $\overline{u'^2}$ is obtained for a single velocity component, say u_1, so this represents the turbulent intensity of one component. If we calculate the mean square value for each component the total energy, known as the turbulent kinetic energy ("tke") is:

$$tke = \frac{1}{2}\left(\left(\overline{u'_1{}^2}\right) + \left(\overline{u'_2{}^2}\right) + \left(\overline{u'_3{}^2}\right)\right) \quad (10.1)$$

The greater the tke typically the more intense is the turbulence. This measure of turbulence can be measured at each location within the flow to determine the spatial distribution of turbulence in a time averaged sense.

Reynolds Averaging

The concept of time averaging of the terms in the Navier-Stokes equation after the velocity is decomposed into a mean and time varying component is known as Reynolds averaging. The instantaneous velocity can be expressed as the sum of the fluctuating part and the mean value at any point in the flow as shown above an rewritten here slightly differently:

$$u = (u' + \overline{u}) \quad (10.2)$$

where the primed quantity is the fluctuating value, which is time dependent, and $\overline{u}$ is the time average value. It should be noted that the time average of the fluctuating part is by definition zero. Each of the time dependent terms in the conservation of mass (or continuity equation) and Navier-Stokes equations can be replaced with the right hand side of Eqn. (10.2). Then each term is time averaged. For instance in the convective acceleration term:

$$\overline{u_j \frac{\partial u_i}{\partial x_j}} = \overline{(u_j' + \overline{u_j}) \frac{\partial (u_i' + \overline{u_i})}{\partial x_j}} = \overline{u_j' \frac{\partial u_i'}{\partial x_j}} + \overline{u_j} \frac{\partial \overline{u_i}}{\partial x_j} \quad (10.3)$$

Where the two terms containing products of the mean and fluctuating parts, when time averaged, are zero since the time average of the fluctuations in zero.

The last term in Eqn. (10.3) is merely the convective acceleration using the time average velocities. Notice that the time averaging and the spatial derivatives can be interchanged. The first term on the very right hand side is the time average of the convective acceleration of the fluctuating velocities. This term is not necessarily zero for a term consisting of the product of two or more time varying signals. This is because the average of the products of time varying signals is not equal to the product of the averages of those signals. Let's examine this term a bit closer for incompressible flow we can write.

$$\overline{u_j{}' \frac{\partial u_i{}'}{\partial x_j}} = \frac{\partial \overline{(u_i{}' u_j{}')}}{\partial x_j} \tag{10.4}$$

Where, for incompressible flow, we are able to bring the velocity u_j inside the derivative with respect to x_j from the continuity condition for incompressible flows. This term can be interpreted as the spatial derivative of the correlation function $\left(\overline{u_i{}' u_j{}'}\right)$ that is the time averaged product of the various velocity components. This is a second order tensor with six independent components since it is a symmetric tensor. This quantity, $\left(\overline{u_i{}' u_j{}'}\right)$ represents the contribution of turbulence to the mean momentum of the flow. That is to say, we have found a term that can change or adjust the mean momentum distribution in the flow caused by averaging the Reynolds decomposition of convective acceleration terms. We will see how this term affects the wall shear stress acting on the fluid for turbulent flow conditions.

Based on our earlier introduction of the turbulent kinetic energy, tke, one might expect that increased levels of tke would be associated with an increased magnitude of $\left(\overline{u_i{}' u_j{}'}\right)$ and increase the effect on the mean momentum distribution. How this occurs and under what conditions it occurs is not a simply thing to determine. This is because the relationship of the energy content of the fluid fluctuations with the momentum transport properties of the turbulence is a set of complicated interactions. Never the less it is possible to observe trends of these interactions and to formulate models of how they work.

We further manipulate the time averaged Navier-Stokes equations by bringing the new fluctuating term $\left(\overline{u_i{}' u_j{}'}\right)$ over to the right hand side of the equation and group it with the viscous stress term:

$$\rho \overline{u_j} \frac{\partial \overline{u_i}}{\partial x_j} = -\frac{\partial \overline{P}}{\partial x_j} + \rho g_i + \left(\frac{\partial}{\partial x_j} \left(\mu \frac{\partial \overline{u_i}}{\partial x_j} - \rho \left(\overline{u_i{}' u_j{}'} \right) \right) \right) \tag{10.5}$$

Here we can see that the viscous term is combined with the momentum altering fluctuating term while all other terms are similar to the laminar flow equation except here we use the time average value of the variables. The last term has two contributions, the mean viscous stress term $\mu \frac{\partial \overline{u_i}}{\partial x_j}$, and the turbulence term. If we interprete the turbulence term as one that modifies the momentum distribution in the flow we can identify it as a "stress" and call it the "turbulent stress", τ_t and write it as:

$$\tau_t = -\rho \left(\overline{u_i{}' u_j{}'} \right) \tag{10.6}$$

In honor of Reynolds, who developed the Reynolds decomposition leading to this results this is often labeled the "Reynolds stress". We can write the total stress within the flow as:

$$\tau_{total} = \tau_{viscous} + \tau_{turbulent} = \mu \frac{\partial \overline{u_i}}{\partial x_j} - \rho \left(\overline{u_i{}' u_j{}'} \right) \tag{10.7}$$

In a number of situations the viscous stress is actually a lot smaller than the turbulent stress and can be ignored except maybe in certain region of the flow, say near a surface or object where the fluctuating quantities are dampened.

Turbulent Boundary Layer

The discussion that follows assumes that the flow is turbulent over the entire surface. The conditions for this are discussed later after this section. Similarly to what was done in laminar flow turbulent boundary layer flows can be analyzed using the same assumptions used to simplify the governing equations. Basically we assume the following for a two dimensional boundary layer flow:

- thin boundary layer, $\delta \ll$ L (or x)
- stream-wise velocity greater than the cross-stream velocity, $u_1 \gg u_2$
- cross-stream derivatives greater than stream-wise derivatives, $\frac{\partial}{\partial x_1} \ll \frac{\partial}{\partial x_2}$
- large Re_{x_1}

The resulting time averaged governing momentum equation becomes for constant property flows and ignoring the body force term:

$$\rho\overline{u_1}\frac{\partial\overline{u_1}}{\partial x_1} \quad \rho\overline{u_2}\frac{\partial\overline{u_1}}{\partial x_2} = -\frac{\partial\overline{P}}{\partial x_1} + \left(\frac{\partial}{\partial x_2}\left(\mu\frac{\partial\overline{u_1}}{\partial x_2} - \rho\left(\overline{u_1{}'u_2{}'}\right)\right)\right) \tag{10.8}$$

Here we have retained the pressure gradient term and only the mean pressure value remains after time averaging the mean plus fluctuating parts.

Notice that in the Reynolds stress term the only contribution is from what is called the "cross correlation", $\overline{\rho u_1{}'u_2{}'}$. This is a consequence of $\frac{\partial}{\partial x_1} \ll \frac{\partial}{\partial x_2}$ even though $\overline{u_1'u_1'}$ may be the same order of magnitude as $\overline{u_1'u_2'}$.

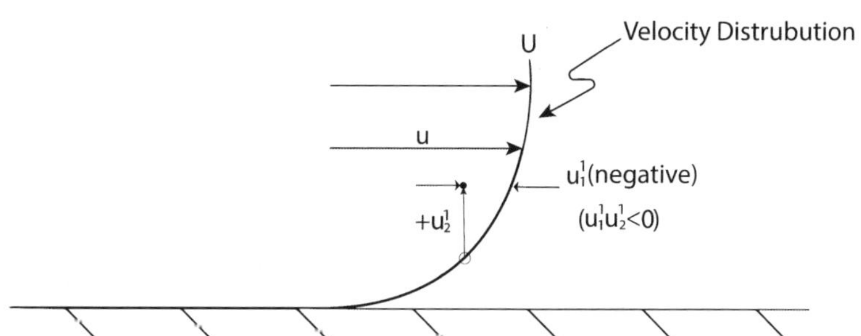

Fig 10.3. Representation of a positive fluctuating velocity in the x_2 direction resulting in a negative value of u_1u_2.

The effect on the mean momentum distribution, caused by turbulence, then is brought about by the cross correlation contained in the Reynolds stress. Referring to Fig. 10.3 we attempt to provide some rationale for how this correlation may impact the momentum distribution and consequently the wall shear stress. At an instant in time imagine a positive, upward value of u_2. This moves fluid with a lower value of u_1 into a region of higher u_1 on average. Therefore the instantaneous value of $u_1{}'u_2{}'$ will be a negative number. Similarly if there is a negative value of u_2 it will move high u_1 velocity fluid into a slower moving region again causing $u_1{}'u_2{}'$ to be a negative number. Assuming this to be the case, one may assume that on average the value of $\overline{u_1{}'u_2{}'}$ is a negative number which means the added turbulent stress is positive, see Eqn. (10.6). This added stress slows down higher speed fluid and speeds up lower speed fluid. The result is a much flatter velocity or momentum distribution across the flow. However, the no-slip condition still holds true. This implies that on average the flatter velocity profile will have higher velocity gradients near the surface, such that the overall frictional force is larger. The net result is that a turbulent boundary layer is expected to have larger surface viscous stresses.

Scaling for Turbulent Boundary Layer Flows

We will define the displacement thickness, δ_1, and the momentum thickness, δ_2, identically as before for laminar flow except we use the time average values of velocity. However, we need to determine an appropriate method to scale the velocity to arrive at an accurate velocity distribution. As for laminar flow we can define the nondimensional cross-stream coordinate as: $\eta = \frac{x_2}{\delta}$. It turns out that the freestream velocity that was used to scale the stream-wise velocity in laminar flow may not be the best scale for the velocity within the boundary layer. The fairly flat velocity distribution of $\overline{u_1}$ in turbulent flow is heavily influenced by the velocity gradient near the wall that determines the wall shear stress. Consequently, a new velocity scale can be introduced that is proportional to the wall stress, called the "friction velocity":

$$v_f = \sqrt{\frac{\tau_s}{\rho}} \tag{10.9}$$

Notice that it is not a specific velocity within the flow but has units of velocity and is proportional to the velocity gradient at the surface through the value of the wall shear stress, τ_s. Since it is reasonable to note that the surface stress depends on the freestream velocity, one can also anticipate that $v_f = f(U)$.

From this we form the nondimensional velocity components:

$$u_1{}^{+} = \frac{\overline{u_1}}{v_f} \tag{10.10a}$$

$$u_2{}^{+} = \frac{\overline{u_2}}{v_f} \tag{10.10b}$$

As we will see in the next section, this scaling of velocity can be used to arrive at an accurate near wall velocity distribution in non-dimensional variables. However, if we choose to try to determine the wall shear stress using an integral analysis similar to what was done in laminar flow, then we can use the following. First we assume a functional form:

$$\frac{\overline{u_1}}{U} = f(\eta) \tag{10.11}$$

In contrast to the polynomial distribution we used in laminar flow. We assume a power law instead of the polynomial (since it has been shown to match experimental data well) as:

$$\frac{\overline{u_1}}{U} = C\eta^{1/n} \tag{10.12}$$

where n is a positive integer and C is a constant. However, if we note that $\dfrac{\overline{u_1}}{U} = 1$ at η =1 then C = 1. Using this expression in the definitions for δ_1 and δ_2 we get:

$$\frac{\delta_1}{\delta} = \frac{1}{n+1} \tag{10.13}$$

$$\frac{\delta_2}{\delta} = \frac{n}{(n+1)\,(n+2)} \tag{10.14}$$

Experimental data shows that n is a weak function of Reynolds number, such that as Reynolds number increases the velocity profile becomes flatter and n increases. For moderately large Reynolds numbers n $\sim$ 7 matches data very well.

To go any further with this we need a relationship between τ_s and δ, similar to what was done with laminar flow, based on the nondimensional velocity distribution. For this we use an alternative scaling with a power law assumption given by:

$$u_1{}^{+} = f_2(\frac{x_2 v_f}{\nu}) = C_2\left(\frac{x_2 v_f}{\nu}\right)^{1/n} \tag{10.15}$$

where C_2 is a constant and $\dfrac{\nu}{v_f}$ is a new length scale. At $x_2 = \delta$, we know that $\dfrac{\overline{u_1}}{U} = 1$ so we can write:

$$\frac{U}{v_f} = C_2\left(\frac{\delta v_f}{\nu}\right)^{1/n} \tag{10.16}$$

This provides the relationship between τ_s and δ, since τ_s is contained in v_f, see Eqn. (10.9). After inserting the definition of v_f and a little algebra we obtain:

$$\frac{\tau_s}{\rho U^2} = C_2{}^{-\frac{2n}{n+1}} Re_\delta{}^{-\frac{2}{n+1}} \tag{10.17}$$

where $Re_\delta = \dfrac{U\delta}{\nu}$ which represents a Reynolds number based on the boundary layer thickness.

Now we use the momentum integral equation that was derived for laminar flow, which is also valid for turbulent flow that yields for a flat plate:

$$c_f = 2\frac{\delta_2}{\delta}\frac{d\delta}{dx_1} = \frac{\tau_s}{\frac{1}{2}\rho U^2} \tag{10.18}$$

Noting that $\frac{\delta_2}{\delta}$ is a constant for a given value of n, and eliminating τ_s between Eqn. (10.17) and (10.18) and solving for δ we obtain:

$$\frac{\delta}{x_1} = C_T Re_{x_1}{}^{-\frac{2}{(n+3)}} \tag{10.19}$$

where $C_T = \left[\frac{\left(\frac{n+3}{n+1}\right) C_2^{\frac{-2n}{n+1}}}{\delta_2/\delta}\right]^{\frac{n+1}{n+3}}$

that is fully determined by the constant C_2 and the value of n. Based on experimental data, much of it analyzed by Blasius for turbulent flow results in:

$$C_2 = 8.74 \tag{10.20}$$

Now we have Eqn. (10.19) fully determined so that we can find the derivative $\frac{d\delta}{dx_1}$ and inserts this into the expression for the skin friction, Eqn. (10.18). For the case of n = 7 we have:

$$C_T = 0.37 \tag{10.21}$$

$$\frac{\delta_2}{x_1} = 0.036\ Re_{x_1}{}^{-1/5}$$

$$c_f = 0.0577\ Re_{x_1}{}^{-1/5}$$

The determination of the total viscous drag force acting on the surface of length L and width S is obtained exactly as was done for laminar flow, by integration of the surface shear stress over the area. The results for the case of n = 7 is:

$$C_f == 0.072\ Re_L{}^{-1/5} \tag{10.22}$$

$$F_D = C_f \frac{1}{2}\rho U^2 LS \tag{10.23}$$

Initial Laminar Starting Length

Turbulent flow over a surface in general will NOT exist from the beginning of the surface, x_1 = 0. Experiments haves shown that turbulent flow on a smooth surface occurs when the Reynolds number, Re_{x_1}, is greater than 5×10^5. Since the Reynolds number is identically zero at the leading edge when x_1 = 0, there will be a region near the leading edge where laminar flow occurs, and then further along the surface the flow becomes turbulent if Re_{x_1} is large enough. Turbulent flow can occur from, or very close to, the leading edge if the flow is tripped. This occurs when there may be slight roughness near the leading edge and the freestream velocity is sufficiently large that turbulence occurs. Under these conditions the Eqns. (10.22) and (10.23) are valid over the entire length of the flow. For the conditions when the flow is not tripped the laminar starting region needs to be considered. A sketch of the boundary layer thickness as influenced by the initial laminar starting length and the transition to turbulence is shown in Fig. 10.4.

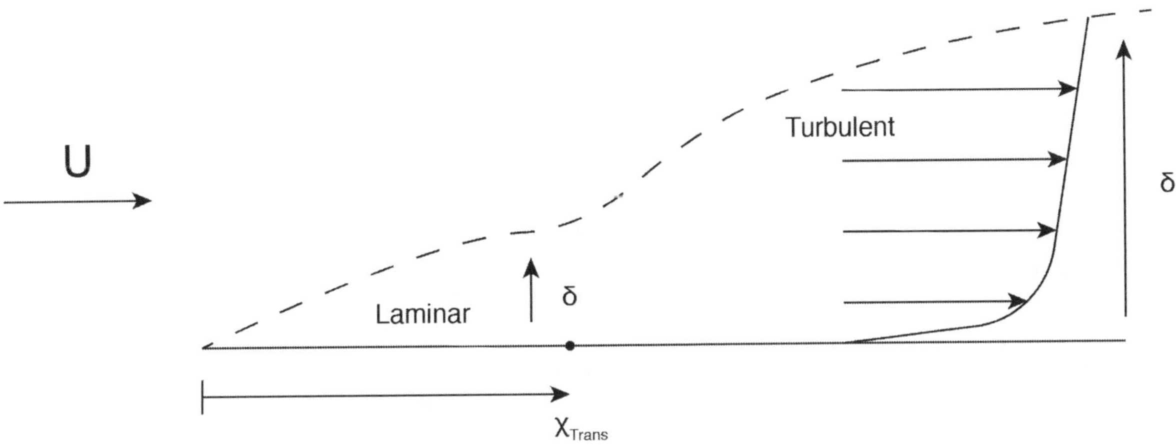

Fig. 10.4: Illustration of the initial laminar starting length and how it influences the boundary layer thickness, transition occurs based on the critical Reynolds number, typically $Re_{crit} = 5x10^5 = Ux_{crit}/v$

The solution to this situation of laminar flow up to the transition location along the surface and turbulent flow beyond this is obtained by dividing up the integration of the surface shear stress into laminar and turbulent regions. The location dividing the two regions is where $x = x_{transition}$ which is found using the critical Reynolds number.

$$x_{transition} = \frac{\left(5x10^5\right)\nu}{U} \tag{10.24}$$

Once the transition location is found then the integration over the entire surface becomes:

$$F_D = \int_o^{x_{transition}} \tau_s dx_1 \; w + \int_{x_{transition}}^{L} \tau_s dx_1 S \tag{10.25}$$

Each term is now divided by $\frac{1}{2}\rho U_\infty{}^2(LS)$ to convert this equation into a nondimensional equation for C_D. The integrand for each term will be the skin friction coefficient, c_f and dx_1 becomes $d(x/L)$. There is a problem here though, the expression for ${c}_{f}$ for the turbulent part (the second integral) can not be solved directly because the skin friction expression found previously assumes the flow begins turbulent at the leading edge. The solution to this problem is to recast the above integral by adding and subtracting the contribution to the total force from $x = 0$ to $x = x_{transition}$ based on turbulent flow. When this is added to the second term in Equation (10.25) the result is an expression for the force assuming turbulent flow from x = 0. The subtraction of this term in the first integral results in the difference between the laminar and turbulent generated force up until $x_{transition}$. Rearranging the terms results in

$$C_D = \int_0^L c_{f,\, turbulent} d\left(\frac{x}{L}\right) - \int_o^{x_{transition}} (c_{f,\, turbuent} - c_{f,\, laminar})\, d\left(\frac{x}{L}\right) \tag{10.26}$$

By using the transitional location given in Eqn. (10.24) the second integral in Eqn. (10.26) can be found directly using the expressions for c_f. Writing this using Re_L (the Reynolds number evaluated at the end of the surface) results in:

$$C_D = \frac{0.072}{Re_L{}^{1/5}} - \frac{1674}{Re_L} \tag{10.27}$$

It should be noted that the above expression assumes n = 7. Also, if transition occurs at a slightly different Reynolds number the coefficient in the numerator of the second term in this equation will vary. Often, the numerator of the second term is 1750 to better fit data.

Universal Velocity Profile

We now introduce the concept of the logarithm velocity profile in the near wall region of a turbulent boundary layer. This concept has proven to match experimental data very well and it is based on the premise that very close to the wall there is a constant shear layer. It also introduces the concept of a turbulent effective viscosity. If we take the stress term in the Navier-Stokes equation, Eqn. (10.7) for a boundary layer as:

$$\tau_{total} = \mu \frac{\partial \overline{u_1}}{\partial x_2} - \rho\left(\overline{u_1{}' u_2{}'}\right) \tag{10.28}$$

We assume that this stress is a constant which can be written in terms of the friction velocity, v_f . Then we model this term based on an effective viscosity, ν_t, that accounts for both the viscous and turbulent contributions which results in the total stress in terms of the friction velocity as:

$$\frac{\tau_{total}}{\rho} = {v_f}^2 = \nu_t \frac{\partial \overline{u_1}}{\partial x_2} \tag{10.29}$$

The effective viscosity (written here as a kinematic viscosity) has units of velocity times length. Taking the friction velocity as the appropriate velocity scale and x_2 as the appropriate length scale (since the viscous effects can only extend away from the wall some amount x_2) then we model the effective viscosity as:

$$\nu_t = \kappa v_f x_2 \tag{10.30}$$

where κ is an unknown constant. Inserting Eqn. (10.30) into (10.29) we have:

$${v_f}^2 = \kappa v_f x_2 \frac{\partial \overline{u_1}}{\partial x_2} \tag{10.31x}$$

This is now integrated in the x_2 direction for a constant value of v_f to obtain a velocity profile as:

$$\frac{\overline{u_1}}{v_f} = \frac{1}{\kappa} ln\ x_2 + c \tag{10.32}$$

This can easily be transformed into the following:

$$\frac{\overline{u_1}}{v_f} = \frac{1}{\kappa} ln\ \frac{x_2 v_f}{\nu} + B \tag{10.33}$$

where κ and B are constants determined from experimental results and shown to be $\kappa = 0.41$ and B = 5. Actually there are also theoretical arguments that can be made to show that $\kappa = 0.41$. This constant stress layer result is only valid fairly close to the surface, but has been shown over a wide range of conditions (different fluids, different Reynolds numbers, etc.) to be very accurate. The non-dimensional variable, $\frac{x_2 v_f}{\nu}$, is often given the symbol ${x_2}^+\ or\ y^+$, and $\frac{\overline{u_1}}{v_f} = u^+$, resulting in a non-dimensional velocity profile valid close to the wall that is expressed in terms of the wall shear stress contained in the variable u^t and x_2^t.

Index